Springer Handbook of Auditory Research

Series Editors: Richard R. Fay and Arthur N. Popper

SPRINGER HANDBOOK OF AUDITORY RESEARCH

Volume 1: The Mammalian Auditory Pathway: Neuroanatomy
Edited by Douglas B. Webster, Arthur N. Popper, and Richard R. Fay

Volume 2: The Mammalian Auditory Pathway: Neurophysiology
Edited by Arthur N. Popper and Richard R. Fay

Volume 3: Human Psychophysics
Edited by William A. Yost, Arthur N. Popper, and Richard R. Fay

Forthcoming volumes (partial list)

Comparative Mammalian Hearing
Edited by Arthur N. Popper and Richard R. Fay

Development of the Auditory System
Edited by Edwin Rubel, Arthur N. Popper, and Richard R. Fay

Clinical Aspects of Hearing
Edited by Thomas R. Van De Water, Arthur N. Popper, and Richard R. Fay

Auditory Computation
Edited by Harold L. Hawkins, Theresa A. McMullen, Arthur N. Popper, and Richard R. Fay

Echolocation in Bats
Edited by Arthur N. Popper and Richard R. Fay

William A. Yost   Arthur N. Popper
Richard R. Fay
Editors

Human Psychophysics

With 58 Illustrations

Springer-Verlag
New York Berlin Heidelberg London Paris
Tokyo Hong Kong Barcelona Budapest

William A. Yost
Parmly Hearing Institute
Loyola University of Chicago
Chicago, IL 60626, USA

Arthur N. Popper
Department of Zoology
University of Maryland
College Park, MD 20742, USA

Richard R. Fay
Parmly Hearing Institute and
Department of Psychology
Loyola University of Chicago
Chicago, IL 60626, USA

Series Editors: Richard R. Fay and Arthur N. Popper

Cover illustration: Detail from Fig. 5.2, p. 161. Measured interaural time differences plotted as a function of source azimuth and elevation.

Library of Congress Cataloging-in-Publication Data
Human psychophysics / William A. Yost, Arthur N. Popper, Richard R. Fay, editors.
    p.   cm. — (Springer handbook of auditory research ; v. 3)
  Includes bibliographical references and index.
  ISBN 0-387-97840-2
  1. Hearing. 2. Psychophysics. I. Popper, Arthur N. II. Yost, William A. III. Fay, Richard R. IV. Series.
BF251.H86 1993
152.1'5—dc20   93-4695

Printed on acid-free paper.

© 1993 Springer-Verlag New York, Inc.
All rights reserved. This work may not be translated or copied in whole or in part without the written permission of the publisher (Springer-Verlag New York, Inc., 175 Fifth Avenue, New York, NY 10010, USA), except for brief excerpts in connection with reviews or scholarly analysis. Use in connection with any form of information storage and retrieval, electronic adaptation, computer software, or by similar or dissimilar methodology now known or hereafter developed is forbidden.
The use of general descriptive names, trade names, trademarks, etc., in this publication, even if the former are not especially identified, is not to be taken as a sign that such names, as understood by the Trade Marks and Merchandise Marks Act, may accordingly be used freely by anyone.

Production managed by Terry Kornak; manufacturing supervised by Jacqui Ashri.
Typeset by Asco Trade Typesetting Ltd., Hong Kong.
Printed and bound by Edwards Brothers, Ann Arbor, MI.
Printed in the United States of America.

9 8 7 6 5 4 3 2 1

ISBN 0-387-97840-2 Springer-Verlag New York Berlin Heidelberg
ISBN 3-540-97840-2 Springer-Verlag Berlin Heidelberg New York

Series Preface

The *Springer Handbook of Auditory Research* presents a series of comprehensive and synthetic reviews of the fundamental topics in modern auditory research. The volumes are aimed at all individuals with interests in hearing research including advanced graduate students, postdoctoral researchers, and clinical investigators. The volumes are intended to introduce new investigators to important aspects of hearing science and to help established investigators to understand better the fundamental theories and data in fields of hearing that they may not normally follow closely.

Each volume is intended to present a particular topic comprehensively, and each chapter will serve as a synthetic overview and guide to the literature. As such, the chapters present neither exhaustive data reviews nor original research that has not yet appeared in peer-reviewed journals. The volumes focus on topics that have developed a solid data and conceptual foundation rather than on those for which a literature is only beginning to develop. New research areas will be covered on a timely basis in the series as they begin to mature.

Each volume in the series consists of five to eight substantial chapters on a particular topic. In some cases, the topics will be ones of traditional interest for which there is a substantial body of data and theory, such as auditory neuroanatomy (Vol. 1) and neurophysiology (Vol. 2). Other volumes in the series will deal with topics that have begun to mature more recently, such as development, plasticity, and computational models of neural processing. In many cases, the series editors will be joined by a co-editor having special expertise in the topic of the volume.

Richard R. Fay
Arthur N. Popper

Preface

Books covering the topics of human psychophysics are usually either textbooks intended for beginning undergraduate or graduate students or review books covering specialty topics intended for a sophisticated audience. This volume is intended to cover the basic facts and theories of the major topics in human psychophysics in a way useful to advanced graduate and postdoctoral students, to our colleagues in other subdisciplines of audition, and to others working in related areas of the neural, behavioral, and communication sciences.

Chapters 2 to 4 cover the basic facts and theories about the effects of intensity, frequency, and time variables on the detection, discrimination, and perception of simple sounds. The perception of sound source location is the topic of Chapter 5. Chapters 2 to 5, therefore, describe the classical psychophysical consequences of the auditory system's ability to process the basic attributes of sound. Hearing, however, involves more than just determining the intensity, frequency, and temporal characteristics of sounds arriving at one or both ears. Chapter 6 argues that perceiving or determining the sources of sound, especially in multisource environments, is a fundamental aspect of hearing. Sound source determination involves the additional processing of neural representations of the basic sound attributes and, as such, constitutes a major component of auditory perception. Chapter 1 provides an integrated overview of these topics in the context of the classical definition of psychophysics.

This volume, coupled with Volumes 1 and 2 of this series covering the anatomy and physiology of the auditory system, should provide the serious student a thorough introduction to the basics of auditory processing. These volumes should allow the interested reader to fully appreciate the material to be covered in future volumes of the series, including those on the cochlea, animal psychophysics, development, plasticity, neural computation, and hearing by specialized mammals and nonmammals.

We are pleased that some of the world's best hearing scientists consented to work with us to produce this volume. We are indebted to them for the time and effort they devoted to writing their chapters. We are also grateful to the staff of Springer-Verlag for enthusiastically supporting the production of this volume.

William A. Yost
Arthur N. Popper
Richard R. Fay

Contents

Series Preface		v
Preface		vii
Contributors		xi
Chapter 1	Overview: Psychoacoustics WILLIAM A. YOST	1
Chapter 2	Auditory Intensity Discrimination DAVID M. GREEN	13
Chapter 3	Frequency Analysis and Pitch Perception BRIAN C.J. MOORE	56
Chapter 4	Time Analysis NEAL F. VIEMEISTER AND CHRISTOPHER J. PLACK	116
Chapter 5	Sound Localization FREDERIC L. WIGHTMAN AND DORIS J. KISTLER	155
Chapter 6	Auditory Perception WILLIAM A. YOST AND STANLEY SHEFT	193
Index		237

Contributors

David M. Green
Psychoacoustic Laboratory, Psychology Department, University of Florida, Gainesville, FL 32611, USA

Doris J. Kistler
Waisman Center, University of Wisconsin, Madison, WI 53706, USA

Brian C.J. Moore
Department of Experimental Psychology, University of Cambridge, Cambridge CB2 3EB, UK

Christopher J. Plack
Department of Psychology, University of Minnesota, Minneapolis, MN 55455, USA

Stanley Sheft
Parmly Hearing Institute, Loyola University of Chicago, Chicago, IL 60626, USA

Neal F. Viemeister
Department of Psychology, University of Minnesota, Minneapolis, MN 55455, USA

Frederic L. Wightman
Waisman Center and Department of Psychology, University of Wisconsin, Madison, WI 53706, USA

William A. Yost
Parmly Hearing Institute, Loyola University of Chicago, Chicago, IL 60626, USA

1
Overview: Psychoacoustics

WILLIAM A. YOST

1. Psychophysics and Psychoacoustics

Hearing is a primary means for us to acquire information about the world in which we live. When someone responds to sound, we say they hear; thus, hearing has two key components: sound and a behavioral response to sound. Psychophysics has been defined as the study of the relationship between the physical properties of sensory stimuli and the behavior this stimulation evokes. Psychoacoustics is the study of the behavioral consequences of sound stimulation, that is, hearing. Psychophysicists, in general, and psychoacousticians, in particular, have searched for functional relationships between measures of behavior and the physical properties of sensory stimulation; for psychoacousticians, this is the search for:

$$\Omega = f(S) \qquad (1)$$

where Ω is a measure of behavior, S is a physical property of sound, and $f(\ )$ represents a functional relationship.

Detection, discrimination, identification, and judging (scaling) have been the primary measures of Ω studied by psychoacousticians. Hearing scientists have described the physics of sound (S) in terms of the physical attributes of the time-pressure waveform or in terms of the amplitude and phase spectra resulting from the Fourier transform of the time–pressure waveform. The functional relationships [$f(\ )$] have ranged from simple descriptions to complex models of auditory processing. However, most of the relationships that exist in the literature have been derived from linear system analysis borrowing from electrical and acoustical descriptions; statistical and probabilistic decision making; and, more recently, knowledge of the neural processing of sound by the auditory nervous system, especially by the auditory periphery.

The basic physical attributes of sound are intensity, frequency, and time/phase characteristics. Simple sounds, which allow the psychoacoustician to carefully study only one of these physical attributes at a time, have been the primary stimuli studied by psychoacousticians. Thus, most knowledge of hearing has been based on sine waves, clicks, and noise stimuli; for these simple stimuli, the field of psychoacoustics has described the detection, discrimination, identification, and scaling of intensity, frequency, and time/phase. Most often, this description has been in the form of a functional relationship which has been couched in terms of linear system analysis.

The sound from any source can be described in terms of its time–pressure waveform or its Fourier transform. Hearing involves a behavioral response to the physical attributes of the source's sound, as described above. However, we can also determine the spatial location of the sound's source based on the sound it produces. The physical variables of sound do not have spatial dimensions, but we can use these variables to determine the spatial dimensions of the source of the sound. Thus, sound localization is another behavioral response (Ω) and psychoacousticians have investigated the relationships between localization and the physical attributes of sound: intensity, frequency, and time/phase. The physical attributes for sound localization have been described as interaural time and intensity differences because, for over 100 years, it has been known that the differences in intensity and phase/timing at the two ears are the primary physical cues used for sound localization. However, as Wightman and Kistler (Chapter 5) show, there are additional physical attributes that affect sound localization.

Developing methods and procedures to determine the functional relationships described by Equation 1 has been a major focus for psychophysics for over 100 years. Psychoacoustics has played a large role in these developments. There are two classes of basic psychophysical procedures: scaling and detection discrimination. The work of Stevens (1975) on scaling and the Michigan group of Birdsall, Green, Swets, and Tanner (Green and Swets 1974) on the theory of signal detection (TSD) have produced both far-reaching methodological contributions as well as theoretical ones for psychoacoustics. Both the refinements of the methods and the continuation of theory development continue to today. For instance, psychoacousticians are still trying to relate measures of loudness, as determined from scaling procedures, to measures of intensity discrimination. Most complete models of auditory processing include "decision stages" which are derived directly from the early TSD work. The forms of the functional relationships that have been formulated have been strongly influenced by these psychophysical methods and theories.

2. Psychoacoustics and the Neural Code for Sound

Let's examine more closely what it means to "hear" in our everyday lives. Hearing usually means determining the sources of sound: you hear the car brake, a person talk, the piano strings vibrating, the wind blowing the leaves, etc. (see Yost 1992). Consider the schematic diagram in Figure 1.1: five potential sound sources are shown and, if all five were producing sound simultaneously, it would be possible to determine the five sources based only on the sounds each produced. However, the sounds from each source do not arrive at our auditory systems as separate sounds. Rather, they are combined

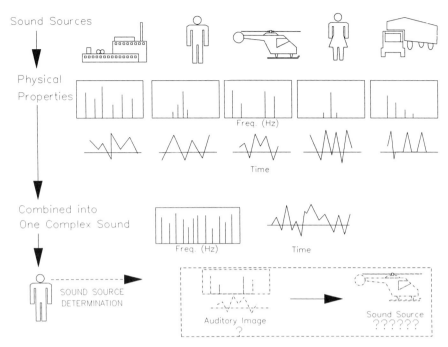

FIGURE 1.1. Five sound sources each produce their own characteristic sound as defined by the time–pressure waveform and/or frequency-domain spectra. The sounds from each source are combined into one complex sound field as the input stimulus for hearing. In order for the auditory system to determine the sources of the sounds (e.g., the helicopter), the physical attributes of the complex sound field must be neurally coded and, then, this neural code must be further processed to allow the listener to determine the sound sources. (Based on a similar figure by Yost 1992.)

into one complex sound field as the sum of the vibratory patterns of each source. That is, although each of the five sources has its own time–pressure waveform, there is only one "stimulus" for hearing. Once this complex input sound stimulus is processed by the auditory system, we can determine that five sources are present and perhaps what and where they are.

The complex sound input must have some neural representation in the auditory system, and this neural representation must contain the information that allows us to determine the sound sources. We now know that the biomechanical and neural elements of the auditory periphery provide the neural code for sound. By establishing the functional relationships between the basic physical attributes of sound and the behavioral measures of hearing such as detection and discrimination, psychoacousticians have described aspects of the neural code of the auditory input as they relate to behavioral responses. Therefore, a great deal of the literature in psychoacoustics, especially in recent years, has been devoted to relating the psychoacoustical indices of neural coding to physiological measures. The collaboration of psychoacoustical and physiological results has produced significant knowledge about the auditory code for the basic physical attributes of sound: intensity, frequency, and time/phase. In addition, this collaboration has produced considerable knowledge about the binaural code for interaural time and intensity differences as the basic stimuli for sound localization.

2.1 Frequency

Since the last part of the nineteenth century, when Helmholtz (see Warren and Warren, 1968, for a translation of some of Holmholtz's work) argued for a resonance theory to explain how the auditory system coded the frequency of sound, the major interest in the hearing sciences has been an explanation of frequency coding. The frequency content of sound does provide the most robust information about a sound source. If we could not determine the frequency content of sound, speech would be Morse code and music would be drum beats. Psychoacoustical data and theories have helped form the current theories for the neural code for frequency. The psychoacoustical conceptions of the critical band by Fletcher (1940) and the excitation pattern by Zwicker (1970) are two cornerstones of the theory of frequency coding. When combined with the cochlear biomechanics of the traveling wave, they form the place theory of frequency coding in terms of channels tuned to frequency, with bandpass filters being the most common method of modeling the tuning of auditory channels for both psychoacoustical and physiological data.

To the extent that the frequency content of sound determines the perceived pitch of the sound, theories of frequency coding also serve as theories of pitch perception. However, as has been known since the days of Helmholtz, a sound's spectral content is not sufficient and necessary to account for a sound's perceived pitch. Studies of the "missing fundamental pitch" (the perception of pitch when the sound has no energy in a frequency region of its spectrum corresponding to the reported pitch) and its many derivatives (virtual or complex pitch; see Terhardt 1974) indicate that a complete theory of pitch perception involves more than a theory of frequency coding. Psychoacousticians have turned to the temporal properties of neural function, especially the fact that neurons discharge in synchrony to the phase of a periodic input, to look for explanations for both pitch perception and to refine models of frequency coding. Moore (Chapter 3) describes in some detail the current state of the psychoacoustical knowledge on frequency processing and theories of pitch perception.

2.2 Intensity

Accounting for the auditory system's sensitivity to the magnitude of sound has proven to be a challenge. The dynamic range for intensity processing is more than 130 dB and, over a considerable portion of this range, changes of

1 dB or less are discriminable. Because a change in a sound's intensity is accompanied by other stimulus changes and because determining if an intensity change has occurred might require memory processes as well as auditory processing, considerable effort has been devoted to carefully determining a human's sensitivity to sound intensity, especially to a change in intensity. The classic ideas of Weber and Fechner, that the just discriminable change in sensory magnitude is a fixed proportion of the intensity being judged (Weber's Law), is not always obtained for sound intensity discrimination (thus, there is a "near miss" to Weber's Law for sound intensity discrimination). Reconciling the fact that Weber's Law is a first approximation to describing intensity discrimination with the facts of the "near miss" to the Law has occupied the efforts of a number of psychoacousticians over the years.

Because sounds rarely exist as isolated sources, measures of intensity processing have often been studied for stimulus conditions involving two sound sources: the signal sound and the background sound. In some conditions, the background sound increases the threshold for detecting the signal sound; thus, the background becomes a masker. Under these conditions, the measure of intensity processing also involves measures of masking.

Optimum signal processing models (most derived from the early work on TSD) have succeeded in accounting for the results obtained in some of the studies summarized above (see Swets, 1964.) The establishment of a neural code for sound intensity has been more difficult to come by. There is agreement that the code must involve the magnitude of the neural responses, but the limited dynamic range of neuronal discharge rate (approximately 40 dB for any one auditory nerve fiber) as compared to the 130 dB dynamic range of hearing is a factual limitation that current theories of neural coding of sound intensity have not overcome.

Recent research on intensity processing and coding has turned to how the auditory system judges a change intensity in one spectral region relative to the intensity of frequency components in other spectral regions. The sounds from most sound sources are characterized by a profile of intensity changes across the sound's spectrum. This spectral profile of a source's sound is unique to that source and does not change when, for instance, the overall level of the sound changes, even over a large intensity range. Humans are remarkably sensitive to changes in the spectral profile of sound that undergoes large overall intensity alterations. Obtaining a better understanding of this sensitivity is crucial if we are to continue to develop the code for processing sound intensity. Green (Chapter 2) reviews many of these findings, especially spectral profile analysis.

2.3 *Temporal Aspects of Sound*

When considering the dimension of time in auditory processing, both the temporal aspects of the stimulus and of the auditory system must be considered. Sounds vary in overall duration and in ongoing changes in level and frequency. A general way of describing time changes for sound is to describe them as amplitude modulated, frequency modulated, or some combination of these two forms of modulation. For instance, a sound that comes on and then goes off $\frac{1}{2}$ second later can be described as a 100% amplitude-modulated sound with one period of a 1-Hz rate of modulation. Or a formant transition in speech can be described as a frequency modulation. Amplitude and frequency modulation may also be applicable for describing stimulus situations that are usually described in terms of temporal masking. For instance, a forward masking paradigm involving a 1000-Hz masker followed by a 1000-Hz signal can be described in terms of the amplitude-modulated envelope of the two sounds. Although this form of description has not often been used in the past, recent theoretical attempts at integrating our knowledge about temporal processing suggest that such descriptions might prove useful. However, to date, no one model, or even a small set of models, has been able to capture the wide range of sensitivity that humans display to the many stimulus conditions that reflect the temporal changes a sound source's waveform may undergo.

The temporal aspects of the hypothesized way in which temporal modulations are processed must also be considered in modeling sensitivity to time or in determining a neural code for time. The nervous system requires time to process the stimulus (e.g., if the system is viewed as a filter, a filter has a time constant) and each neural element will have temporal constraints (e.g., a refractory period). The neural code for time can range from the simple description that the duration of some neural event represents the duration of the stimulus to the more complicated formulations based on the temporal pattern of neural discharges coding for some aspect of the temporal change that the stimulus has undergone.

Models of temporal integration (integrating information over time until sufficient information has been achieved resulting in a response) have attempted to account for the effect of overall duration on auditory sensitivity. No one integration process has been able to account for the multitude of stimulus conditions that have been tested, especially when the conditions involve ongoing temporal changes. However, recent efforts to provide a general account of stimuli that can be described as amplitude modulated in one way or another are proving more successful than past attempts. Less success has been achieved for stimuli containing frequency modulation. These and other aspects of temporal processing are reviewed by Viemeister and Plack (Chapter 3).

2.4 Localization

For over a hundred years, the interaural differences of time and intensity have been known to be crucial variables in determining the location of a sound in the horizontal or azimuth plane. In the early years of the study of localization, the processing of these variables was characterized by the duplex theory of localization (see Yost and Hafter 1987) which stated that low-frequency sounds were localized on the basis of interaural time differences and high-frequency sounds on the basis of interaural intensity differences. Jeffress (1948) first proposed a neural coincidence network as a possible means for computing the cross-correlation between sounds arriving at the two ears as a means of coding the interaural time difference as it would be used to aid localization. Additional modeling and physiological work has continued to support cross-correlation and coincidence networks as crucial elements in coding both interaural time and intensity differences.

A number of findings over the past 20 years have led psychoacousticians to modify significantly the duplex theory of localization. The stimulus variables responsible for vertical localization are probably not the same as those used for azimuth localization, and an even different set of variables probably govern our ability to localize a source based on range or depth. High-frequency sounds can be localized on the basis of interaural time differences, if the sounds have a slow temporal modulation in their temporal envelopes. There are situations in which subjects have been shown to localize sound with presumably only monaural information. These and other findings have moved psychoacousticians away from the duplex theory.

The newest development in understanding the code for localization stems from demonstrations that the transformations that sound undergoes as it travels from the source across the head and torso of a person to the two ears are crucial for establishing our sense of auditory space. The head-related transfer functions (HRTFs) describe these transformations and they have been measured with great care for a number of stimulus conditions. By using the information in the HRTFs, an age-old problem in binaural hearing appears on the verge of an answer. When stimuli are presented over headphones and the two ears are stimulated with differences of time and/or intensity, subjects report a perceived location for the stimuli, but the sources appear "inside the head" rather than "out" in the external environment. Thus, psychoacousticians developed the nomenclature of lateralization to describe experiments involving headphone presentations of sounds and localization for nonheadphone presentations. Because of the relative ease of conducting headphone studies and the stimulus control such studies afforded, most of the data concerning binaural hearing have been lateralization data. However, until the HRTFs were taken into account when presenting stimuli to listeners over headphones, few headphone studies could establish a perception of sound sources occurring in real space as they do in the normal, everyday process of localization. Now, by using digital signal processing tools and knowledge about HRTFs, it appears possible to use the ease and stimulus control made possible with headphone-delivered stimuli to study localization as it occurs in real environments. These and other important issues of binaural hearing are reviewed by Wightman and Kistler (Chapter 5).

The studies of frequency, intensity, temporal, and binaural processing as described above have provided a wealth of data and theories on the sensitivity of humans to these basic variables of sound. That is, psychoacoustics can formulate a number of valid and reliable functional relationships between behavioral measures of detection, discrimination, identification, scaling, and localization with the physical variables of frequency, intensity, time, and interaural time and intensity differences. There are, therefore, a number of solutions for Equation 1.

2.5 Auditory Perception

Returning to Figure 1, the history, summarized in Section 2, has a lot more to say about how the complex sound field for any real world collection of sound sources is coded in the auditory system than it does about how the individual sources are determined. It appears that the neural code for the sound field must undergo additional processing in order for the auditory system to determine the sources of the sounds that make up the sound field, especially when more than one sound source is present.

Let us consider a simple example of two sound sources, each of which contains three frequencies. Sound source A is composed of frequencies F_i, F_k, and F_m whereas sound source B is composed of frequencies F_j, F_l, and F_n. As implied by the subscripts, the spectral components overlap and it will be assumed that the two sound sources are presented simultaneously. Thus, a complex sound field, consisting of six frequencies ($F_i, \ldots F_n$), forms the auditory stimulus. A basic question for hearing is: How are the three components of source A and the three of source B represented in the auditory system to allow for the determination of two sound sources? Some physical attributes of the components of source A must be similar in value but distinct from those of the three components of source B. For instance, all the components of source A may share a common value of some physical attribute (e.g., intensity) which is different from the value shared by the three components of source B. It is assumed that the six component frequencies are each neurally coded by the auditory system. The auditory system must recognize that all those neurally coded components that share a common attribute form one neural subset as the basis for determining one sound source, that is, distinct from other similarly formed neural subsets representing other sound sources.

Although the auditory system's remarkable acuity for locating sound sources is probably a crucial aspect of our ability to determine sound sources, variables other than just spatial location allow for sound source determination. For instance, listeners experience little, if any, difficulty in hearing the full orchestra and most of its constituent instruments with monaurally recorded and played back music. In fact, in terms of identifying the music and the instruments (the sound sources), listening to the monaural recording is only marginally poorer than being present at the actual concert. Put another way, determining the location of a sound source does not necessarily reveal

TABLE 1. Four questions pertaining to auditory processing for determining the sources of sound

1. What physical attributes of sound sources are used by the auditory system to determine the sources?
2. How are those physical attributes coded in the nervous system?
3. How are the neurally coded attributes associated with any one sound source pooled or fused within the auditory system such that there is a neural subset representing that sound source?
4. How are the various neurally coded pools of information segregated so that more than one sound source can be determined?

Adapted from Yost 1992.

what the source is. This discussion suggests at least four questions (see Table 1) that require answers if we are to understand how the sources of sounds are determined.

Determining the sources of sound depends on the auditory system using the information available in the neural code for sound. Auditory perception can be defined as the processing of this sensory code for sound. Thus, sound source determination has become a major topic of auditory perception (see Bregman 1990 for an explicit discussion and Handel 1989 for an implicit discussion of this observation). Prior to the recent focus on sound source determination, speech and music perception were often cited as the areas of auditory perception. To some, auditory perception as sound source determination can encompass special sounds such as speech and music (Bregman 1990). To others, the perception of speech and music requires special perceptual mechanisms apart from or in addition to those used to determine sound sources (Liberman and Mattingly 1989). Thus, there is currently no unified view of auditory perception.

Consideration of the processing of sound that might be necessary to determine sound sources requires modifications in the basic formulation for establishing psychoacoustical relationships as described in Equation 1. First, for many questions related to sound processing, the neural code for the physical variable of sound becomes a more appropriate independent variable in the equation than just the stimulus value (S). That is, it might be crucial to know how a particular physical parameter of a sound source is represented in the nervous system rather than only what that physical value may be. Second, questions of sound source determination often deal with interactions among many components across the spectrum and/or time rather than dealing with one or a small number of variables at a time. That is, when there is more than one sound source, the auditory system will probably have to process information across both the spectrum of the incoming sound field as well as over time to determine the sources. And finally, there might have to be a refinement in the behavioral measures and procedures used to determine how humans process information about the sources. The current procedures for measuring detection, discrimination, identification, scaling, and localization might not be sufficient to measure sound source determination. The basic concept conveyed in Equation 1 still holds as psychoacoustics attempts to determine functional relationships between behavior and sound. However, the details concerning each part of this relationship are changing as psychoacoustics moves from questions pertaining to descriptions of the sensitivity of the auditory system to the basic properties of the sound field required to provide a neural code for the sound field, to questions concerning more complex stimuli and interactions that occur when the auditory system must process the information in the sound field to determine sound sources. Yost and Sheft (Chapter 6) describe some of the work that has and is being done to understand auditory perception and sound source determination.

3. An Example

Figure 1.2 shows the time-domain outputs (over 200 ms) of a bank of critical-band filters (Patterson et al. 1992) stimulated with seven tones from two theoretical sound sources: one source consisting of four steady-state tones

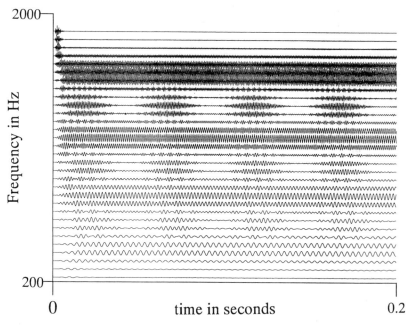

FIGURE 1.2. The output of a simulation (Patterson et al. 1992) of the processing provided by the auditory periphery to a seven-tone complex consisting of four tones presented without amplitude modulation and three tones that are amplitude modulated at a 20-Hz rate. The time–domain outputs of 31 tuned auditory channels are shown for the first 200 ms of stimulation. The channels are spaced at intervals of one equivalent rectangular bandwidth (ERB). (See Moore and Glasberg 1987; Moore, Chapter 3.)

(having frequencies of 300, 520, 875, and 1450 Hz) and the other source consisting of three amplitude-modulated (at a rate of 20 Hz) tones (having frequencies of 400, 675, and 1125 Hz). This example characterizes some of what has been learned about hearing from psychoacoustical research as well as some of the questions that remain to be answered.

The outputs of these filters represent a realistic estimate of the vibratory information that the hair cells process. The characteristics of these filters and their spacing is the result of years of psychoacoustical research on masking and frequency selectivity. The fact that the time-domain outputs of these filters are reasonable estimates of neural input is reinforced by psychoacoustical data on temporal envelope processing. These and additional psychoacoustical data and theories have enabled a number of researchers (e.g., Patterson et al. 1992) to implement models of auditory processing that provide a very realistic description of the auditory code for sound. Recent psychoacoustic work has also shown that the three-tone amplitude modulated (AM) complex is perceived separately from the four-tone steady-state portion of the spectrum, that is, the AM complex is segregated from the four-tone complex as if there were two sound sources.

It is clear that, according to this model of auditory coding, the amplitude modulation of the two sources is preserved in the neural code for these two sound sources. By looking at the time-domain waveforms from each filter, it is not difficult to predict that there might be two sound sources. Thus, it is clear that this level of modeling can describe the neural characteristics of the tonal components that might enable the nervous system to determine the existence of two sound sources. However, just because we can "see" the pattern of activity does not mean we understand how the nervous system processes the amplitude modulation differences to aid the organism in determining that two sources produced this complex sound. Little is known about how amplitude modulation is processed to allow for the determination of two sources. However, the recent psychoacoustical literature (see Yost and Sheft, Chapter 6) documents the beginning attempts to gain this knowledge. Therefore, psychoacoustics has formed and continues to form functional relationships between the physical variables of sound and the behaviors of detection, discrimination, recognition, scaling, and localization. Psychoacousticians are beginning to seek similar relationships between these physical variables and our abilities to determine the sources of sounds in our environment.

References

Bregman AS (1990) Auditory Scene Analysis. Cambridge, MA: MIT Press.
Fechner GT (1860) Elemente dr Psychophysik. Leipzig, Germany: Breitkopf u. Hartel.
Fletcher H (1940) Auditory patterns. Rev Mod Phys 12:47–65.

Green DM, Swets JA (1974) Signal Detection Theory and Psychophysics. New York: Robert E. Krieger.

Liberman AL, Mattingly IG (1989) A specialization for speech and perception. Science 243:489-494.

Moore BCJ, Glasburg BR (1987) Formulae describing frequency selectivity as a function of frequency and level, and their use in calculating excitation patterns. Hear Res 28:209-225.

Patterson RD, Robinson K, Holdsworth J, McKeown D, Zhang C, Allerhand M (1992) Complex sounds and auditory images. In: Cazals Y, Horner K, Demany L (eds) Auditory Physiology and Perception. Oxford: Pergamon Press, pp. 417-429.

Stevens SS (1975) Psychophysics: Introduction to its perceptual, neural, and social aspects. In: Stevens G (ed) New York: John Wiley and Sons.

Swets JA (ed) (1964) Signal Detection and Recognition by Human Observers. New York: John Wiley and Sons.

Terhardt E (1974) Pitch, consonance and harmony. J Acoust Soc Am 55:1061-1069.

Yost WA (1992) Auditory image perception and analysis. Hear Res 56:8-19.

Yost WA, Hafter E (1987) Lateralization of simple stimuli. In: Yost WA, Gourevitch G (eds) Directional Hearing. New York: Springer-Verlag, pp. 49-85.

Warren RM, Warren R (1968) Helmhlotz on Perception: Its Physiology and Development. New York: Wiley.

Zwicker E (1970) Masking and psychological excitation as consequences of the ear's frequency analysis. In: Plomp R, Smoorenburg GF (eds) Frequency Analysis and Periodicity Detection. Leiden, The Netherlands: AW Sijthoff, pp. 393-403.

2
Auditory Intensity Discrimination

DAVID M. GREEN

1. Introduction

This chapter summarizes what is known about the discrimination of sounds that differ in intensity. The topics are: (1) differential intensity limens for pure tones; (2) intensity discrimination tasks in which random noise is used as a masker; and (3) discrimination of intensity changes in complex signals. For each topic, a summary of the principal facts will be provided, as well as a discussion of the most important stimulus parameters and how they influence the experimental outcomes. The emphasis will be on the empirical over the theoretical. Although I will try to indicate the theoretical approaches that have been used to explain certain facts, the theories will not be covered in sufficient detail to permit anything more than a general appreciation of the method of attack.

1.1 History

One might expect that intensity discrimination is a process that is fairly well understood, given the salience of this auditory dimension, but such an expectation is false. As I have commented in a recent monograph (Green 1988), "at practically every level of discourse, there are serious theoretical problems that prevent offering a complete explanation of how the discrimination of a change in intensity is accomplished." Part of this ignorance can be traced to the relatively short time that we have been able to carry out quantitative studies in this area. A major problem has been the lack of stimulus control. This inability to control sound intensity is particularly striking when compared with the control and measurement of sound frequency. We have been able to measure the frequency of a periodic vibration with third-place accuracy for at least two centuries. Control of stimulus frequency was also convenient, since it initially depended only on the availability of stretched strings and, later, on tuning forks. The control and measurement of acoustic intensity was quite a different issue. The actual measurement of sound pressure level dates from 1882 when Lord Rayleigh's theoretical analysis provided the relation between sound pressure level and the torque exerted on a light disc suspended in the sound field (Rayleigh, 1882). It was not until the electronic revolution of the past several decades that a convenient and precise means of controlling the level of the auditory stimulus became available. It is sobering to recall that less than 100 years ago a popular means of controlling auditory intensity was to drop objects at various heights above a sounding block (Pillsbury 1910). Riesz's (1928) study, at the Bell Telephone Laboratories, of the intensive limen for a pure tone was the first investigation to use the new electronics, and it marks the beginning of modern psychoacoustics. In addition to the electronics used to produce the stimulus, the advent of digital computers has also revolutionized psychoacoustics laboratories. In the earlier studies, the usual stimulus was either a pure tone or random noise. Occasionally, investigators used a contrived signal such as a pulse train or a square wave—a signal that contained a number of different frequency components. More frequently, however, the single sinusoid was the stimulus of choice. With the availability of high-speed digital-to-analog converters and inexpensive computer memory, complex acoustic signals comprised of components with independently adjustable phases, frequencies, and amplitudes are limited only by the imagination of the investigator. Complex auditory signals are being studied in a number of different laboratories in a number of different research programs. The results obtained with these more complicated stimuli are often surprising and unexpected. Our knowledge of simple stimuli often cannot be transferred to the results obtained with more complex stimuli.

1.2 *Successive and Simultaneous Comparison*

Before beginning the review of the empirical data, the general process of discriminating between two sound intensities will be considered. There are, in fact, two quite distinct tasks that can be posed to the listener when investigating the discrimination of changes in intensity. The listener's memory requirements are very different in these two tasks, and it is not surprising to find, on that basis alone, that the results obtained with these two tasks are quite different. In the classical approach to intensity discrimination, the sounds are presented *successively* to the listener. In effect, a pair of waveforms, $f(t)$ and $k \cdot f(t)$, are presented where k is a constant greater than one that increases the intensity of the second sound. This discrimination task amounts to determining the order in which the pairs are presented, loud-soft or soft-loud.

To accomplish this discrimination, the minimal requirements are that the listener must estimate the level of two sounds and compare the two estimates. Since the sounds are presented in succession, the listener must retain some memory of the first estimate so that it can be compared with the later estimate. Historically, scant attention has been paid to the memory process. In fact, since the overall level of the stimuli was generally fixed for an extended period of time, the listener probably developed a long-term memory of the "standard" level. As each sound was presented, the estimated level was compared with the "standard" level and, in effect, the listener could make a reasonable estimate about the level of each stimulus as it was presented. The successive nature of sound presentations was, in fact, irrelevant. The importance of this long-term memory of the standard only began to become apparent in the mid-1950s. It was clear that a long-term memory standard was present when Pollack (1955) separated the interstimulus interval by 24 hours and found only a slight deterioration of the discrimination accuracy.

A simple procedure allows the experimenter to prevent the listener from developing this long-term memory. The experimenter can use pairs of sounds to be discriminated and randomly vary the overall level of the pairs. This is known in the jargon as "roving" the stimulus level. The first pair might be presented at 60 and 61 dB SPL and the next pair at 30 and 31 dB SPL*. Because the level changes for each pair, the listener must do exactly what was initially assumed, namely, compare the stimuli successively and make a decision on that basis. When the discrimination experiment uses a roving level, then the interstimulus interval becomes a critical experimental parameter. Tanner (1961), Sorkin (1966), and Berliner and Durlach (1973) all demonstrated that interstimulus intervals (ISI) are important in discrimination experiments. They varied the ISI from less than one to several seconds and produced a noticeably different discrimination performance. In general, the longer the interstimulus delay, the poorer the performance, presumably because the memory of the first sound decayed before the second sound was presented. It should be noted that most of the classical data were obtained with fixed stimulus levels so that, in all probability, the short-term memory factor was not involved in such studies.

A second way to study intensity discrimination is to require the listener to compare at least two sounds that are presented *simultaneously*. Results using this procedure are very recent. They include experiments where the listener tries to detect a change in spectral shape (profile analysis, Green 1988) as well as an experimental paradigm first described by Hall, Haggard, and Fernandes (1984), usually called comodulation masking release (CMR). In these studies, the memory processes were minimal. The listener was asked to compare two spectral regions of a single sound. Often, two sounds were presented: in one, the spectral levels were the same; in the other, they differed. Green, Kidd, and Picardi (1983) have shown that the interstimulus interval has very little effect on the results obtained in such experiments. Their results also showed that simultaneous intensity comparisons were often more sensitive than successive comparisons. In the simultaneous task, they argued that the listener could decide after each presentation whether the two sounds were the same

---

* Sound pressure level (SPL) is the level of the sound in decibel re a specific sound pressure level, namely, 0.0002 dynes/cm^2.

or different. This new area of research will be reviewed after the review of the traditional data.

One troublesome aspect of the traditional approach is that it often assumes that the discrimination process is a successive comparison because the stimuli are presented that way. In some cases, there are strong reasons to suspect that profile analysis (simultaneous comparisons) is operating, and the results have little to do with the successive process. Green (1988), for example, has shown that the detection of a sinusoidal signal in noise, generally analyzed from the viewpoint of energy detection, is essentially unaffected when the overall level of the stimuli is varied on each and every presentation over a 40-dB range. Clearly, an energy detector making successive comparisons must be affected by 40-dB changes in level. Thus, it should be realized that at least some conventional wisdom may be severely challenged by this new appreciation of how precise such simultaneous intensity comparisons can be.

2. Difference Limen for a Pure Tone

2.1 Riesz's Study

The first systematic investigation of auditory intensity discrimination using a sinusoidal signal was that of Robert Riesz at the Bell Telephone Laboratories in 1928. At the time, turning a sinusoidal signal on and off produced noticeable clicks. How these unwanted transients would affect the results of the experiment was unknown, so Riesz used a procedure that avoided the problem. He used two sinusoidal components of nearly the same frequency and very different amplitudes. The result was the familiar beating waveform—a carrier that has a frequency nearly equal to the frequency of the larger amplitude component, with an envelope that is also sinusoidal with a frequency equal to the difference in frequency of the two components. The maximum amplitude of the beating wave-form is equal to the sum of the two component amplitudes. The minimum amplitude is equal to the difference between the larger and smaller amplitudes (see Rayleigh 1877, p. 23). The procedure used to determine the discrimination threshold could almost be considered casual in light of modern procedures. The threshold value was determined by the listeners adjusting the amplitude of the smaller component until they could just hear the fluctuation in amplitude.

Riesz (1928) carried out some preliminary investigations to determine how the sensitivity to such beats was related to the beat frequency. Very slow fluctuations in amplitude were difficult to hear, presumably because of memory limitations. Very high beat frequencies were also difficult to hear, presumably because the amplitude fluctuations occurred so rapidly that they could not be followed. At moderate beat rates, sensitivity to the amplitude changes was maximal. Hence, Riesz used a 3-Hz beat rate in all his measurements. He studied a wide range of stimulus conditions; the carrier frequency ranged from 35 to 10,000 Hz and the intensity levels ranged from near threshold to 100 dB above threshold at some frequencies.

A complication introduced by the use of this procedure is that the stimulus must be continuously present. Later investigators have used a gated presentation procedure: two sinusoids were presented successively in time, with one larger in amplitude than the other. There are several studies which suggest that the gated procedure may produce data somewhat different from those obtained in continuous (modulation) presentation procedures. The general trends of Riesz's data, however, have been replicated in all the more recent studies. Whether any of the differences between Riesz's data and any modern data are due to the continuous versus gated modes of presentation or to other differences in procedure is not fully understood at this time.

2.2 Measures of Intensity Change

An issue still unresolved is what physical measure should be used to summarize the results of intensity discrimination experiments. About all the different investigators can agree upon is that the measure should be dimensionless. In Riesz's study, a simple measure of sensitivity is the relative size of the small and large components. If we denote the larger component as having amplitude p and the smaller as having amplitude Δp, then the ratio $\Delta p/p$ is one obvious means of summarizing the listener's adjustments. The smaller the number, the more sensitive the listener is to changes in amplitude.

The problem is that each investigator feels that he or she knows the "proper" way to measure the stimulus. Thus, the investigator begins with the "correct" analysis of the discrimination task and ends with what he or she considers to be the "natural" measure of the stimulus. For example, in Riesz's experiment, the maximum beat amplitude is actually $(p + \Delta p)$ and the minimum $(p - \Delta p)$, so the max/min ratio is obviously $(p + \Delta p)/(p - \Delta p) = (1 + \Delta p/p)/(1 - \Delta p/p)$, which, if $\Delta p/p$ is small, is approximately $1 + \Delta p/p$. Others might argue that the power or energy of the signal is a more "natural" measure than stimulus pressure. They, therefore, calculate the maximum and minimum of quadratic quantities such as $(p + \Delta p)^2$ and $(p - \Delta p)^2$. Still another set of measures arises from the logarithm of these dimensionless quantities, a decibel-like unit. Grantham and Yost (1982) have presented a balanced discussion of the various measures.

Naturally, this author has his own prejudices, but the student attempting to understand this area must first recognize that there is a variety of measures and that different investigators use different measures. As an eminent psychophysicist once put it, the problem of psychophysics is "the definition of the stimulus" (Stevens 1951). In my opinion, we will be in a position to "correctly" define the stimulus only when we understand a great deal more about intensity discrimination and how it works. At present, we are considerably short of that goal. Therefore, I will summarize the various measures that have been proposed and cite the more obvious approximations among them. These approximations are true only when discrimination accuracy is acute. Unfortunately, most intensity discrimination data fall in a range where the approximations are only marginally correct. Most investigators claim that their favorite measure is the true "Weber fraction," for example, $\Delta p/p$, so any specific definition of that term will be avoided.

There are five quantities most often used to index the threshold for discriminating a change in intensity. They involve ratios of pressures or intensities (powers, energies) and logarithmic quantities derived from them. These definitions depend on the intensity of a sound, I, being proportional to the sound pressure squared, p^2, with ΔI therefore being proportional to $(p + \Delta p)^2 - p^2$.

1. Pressure ratio: $pr = \Delta p/p$
2. Intensity ratio: $\Delta I/I = [(p + \Delta p)^2 - p^2]/p^2 = (2p\Delta p + \Delta p^2)/p^2$

$$\Delta I/I = 2\Delta p/p \text{ if } \Delta p/p \ll 1$$

Logarithms of quantities related to 1 or 2:

TABLE 2.1. Five measures of Weber fraction

$20\log(\Delta p/p)$	$10\log(\Delta I/I)$	ΔL	$\Delta I/I$	$\Delta p/p$
−40	−16.97	0.09	0.020	0.010
−39	−16.47	0.10	0.023	0.011
−38	−15.96	0.11	0.025	0.013
−37	−15.46	0.12	0.028	0.014
−36	−14.96	0.14	0.032	0.016
−35	−14.45	0.15	0.036	0.018
−34	−13.95	0.17	0.040	0.020
−33	−13.44	0.19	0.045	0.022
−32	−12.94	0.22	0.051	0.025
−31	−12.43	0.24	0.057	0.028
−30	−11.92	0.27	0.064	0.032
−29	−11.41	0.30	0.072	0.035
−28	−10.90	0.34	0.081	0.040
−27	−10.39	0.38	0.091	0.045
−26	−9.88	0.42	0.103	0.050
−25	−9.37	0.48	0.116	0.056
−24	−8.85	0.53	0.130	0.063
−23	−8.34	0.59	0.147	0.071
−22	−7.82	0.66	0.165	0.079
−21	−7.30	0.74	0.186	0.089
−20	−6.78	0.83	0.210	0.100
−19	−6.25	0.92	0.237	0.112
−18	−5.72	1.03	0.268	0.126
−17	−5.19	1.15	0.302	0.141
−16	−4.66	1.28	0.342	0.158
−15	−4.12	1.42	0.387	0.178
−14	−3.58	1.58	0.439	0.200

TABLE 2.1 (continued)

$20\log(\Delta p/p)$	$10\log(\Delta I/I)$	ΔL	$\Delta I/I$	$\Delta p/p$
−13	−3.03	1.75	0.498	0.224
−12	−2.48	1.95	0.565	0.251
−11	−1.92	2.16	0.643	0.282
−10	−1.35	2.39	0.732	0.316
−9	−0.78	2.64	0.836	0.355
−8	−0.20	2.91	0.955	0.398
−7	0.39	3.21	1.093	0.447
−6	0.98	3.53	1.254	0.501
−5	1.59	3.88	1.441	0.562
−4	2.20	4.25	1.660	0.631
−3	2.83	4.65	1.917	0.708
−2	3.46	5.08	2.220	0.794
−1	4.11	5.53	2.577	0.891
0	4.77	6.02	3.000	1.000
1	5.44	6.53	3.503	1.122
2	6.13	7.08	4.103	1.259
3	6.83	7.65	4.820	1.413
4	7.54	8.25	5.682	1.585
5	8.27	8.88	6.719	1.778
6	9.02	9.53	7.972	1.995
7	9.77	10.21	9.489	2.239
8	10.54	10.91	11.333	2.512
9	11.33	11.64	13.580	2.818
10	12.13	12.39	16.325	3.162

3. $Lp = 20\log(\Delta p/p)$
4. $LP = 10\log(\Delta I/I)$

Unfortunately these are not simply related:

$$LP = 3 \text{ dB} + Lp/2 \text{ if } \Delta p/p \ll 1$$

Finally, a difference in the level of the two sounds to be discriminated:

5. Level difference in decibels, ΔL

$$\Delta L = 20\log(p + \Delta p) - 20\log p$$
$$= 10\log(p + \Delta p)^2/p^2$$
$$= 10\log(1 + \Delta I/I)$$
$$\Delta L = 4.343(\Delta I/I) = 8.686(\Delta p/p) \text{ if } \Delta p/p \ll 1$$

Table 2.1 lists the relationship among these five quantities. Typical discrimination performance ranges between −20 and −10 dB [$20\log(\Delta p/p)$], so the approximations are not very accurate in this range.

2.3 Psychometric Functions

Implicit in the definition of a threshold value for the stimulus is the psychometric function, the function relating percentage of correct discrimination responses to the change in stimulus intensity. A "threshold value" for intensity discrimination amounts to determining the intensity value corresponding to a particular performance level on the psychometric function. If a two-alternative forced-choice procedure is used, then the proportion of correct responses, $P(c)$, varies from 0.5 to 1.0 as the stimulus level is varied. The choice of a threshold value is completely arbitrary, but the midway value of 0.75 is often used to specify the threshold value of the stimulus. If adaptive procedures are used (Levitt 1971), then the common two-down, one-up procedure tracks the 0.707 proportion, and this value is used to designate the threshold value.

While there are differences in *interpretation*, there is, in fact reasonably good agreement among the different empirical studies of the psychometric function. The disagreements arise because of the different measures used to express the stimulus. Green et al. (1979) and Hanna, von Gierke, and Green (1986) have shown that, in a two-alternative forced-choice task, the psychometric function is reasonably approximated by the equation

$$d' = C \, \Delta p/p \qquad (1)$$

where c is a constant and d' is the argument of the normal or Gaussian distribution function, that is, $P(c) = \Phi(d')$. An example of such a psychometric function and some data can be see in Figure 2.1; it is the solid line on the left in the figure.* This equation is not, however, the only way to describe the psychometric function. The reader can note that for small values of $\Delta p/p$, ΔL is proportional to $\Delta p/p$ and the constant of proportionality is about 8.6. Thus, the psychometric function could also be approximated by the following equation

$$d' = k \, \Delta L \qquad (2)$$

where k is a constant, as Pynn, Braida, and Durlach (1972) and Rabinowitz et al. (1976) have suggested. In those experiments, $\Delta p/p$ was small, and it would be impossible to choose between the two relationships on the basis of the data because of the linearity of ΔL and $\Delta p/p$. In short, there is agreement on the empirical evidence but disagreement over the preferred measure to use on the abscissa.

As will be seen in the next section, there are good theoretical reasons to expect a linear relation between d' and $\Delta p/p$. An ideal detector trying to

---

* The dotted curve is a theoretical fit to date in which the listener was detecting a single sinusoid in noise, rather than discriminating between two sinusoidal intensities. Clearly there is a difference in the shape of the psychometric function for these two tasks.

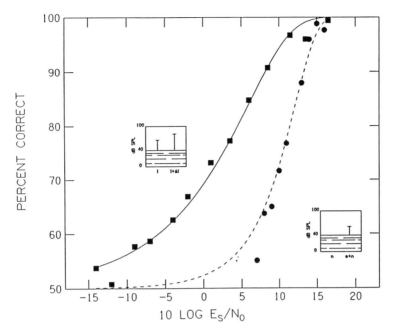

FIGURE 2.1. Psychometric functions for two types of detection tasks. The ordinate is the percentage of correct responses in a two-alternative forced-choice task. The abscissa is the signal energy, E_s, to noise power density (noise power per cycle), N_o. The two square panels schematically represent the experimental situation used to obtain the psychometric functions. For the function on the right, the sinusoidal signal of energy, E, is added to a wide-band masking noise. For the function on the left, two equal-amplitude tones that are clearly audible are presented in both intervals. A signal of energy, E_s, is added, in phase, to one of the tones. The detection task is to select the more intense tone.

detect the change in the amplitude of a sinusoid presented in Gaussian noise would have such a psychometric function. The same formula would apply if it is assumed that a fundamental limitation on discriminating a change in intensity is also some kind of internal, Gaussian fluctuation that can be treated as adding to the stimulus. This is exactly the assumption made by Laming (1986) in his theory. But, again, the same theoretical arguments could be used to predict a linear relation between d' and ΔL.

Buus and Florentine (1991) have marshaled evidence to support the selection of ΔL as the "proper" variable to be used in expressing the psychometric function. One simple way to achieve some measurable differences among the various possible forms of the psychometric function is to measure conditions where the threshold value is very high, $\Delta p/p \gg 1$. In that case, there will be a nonlinear relation between the three major measures, $\Delta p/p$, $\Delta I/I$, and ΔL.

Thus, it may be possible to choose among them. Buus and Florentine used very short signal durations ($T = 10$ ms) and a variety of signal frequencies (250, 1000, 8000, and 14,000 Hz) in their study. They also used longer duration signals to contrast the short- and long-duration psychometric function. They fitted their data to a general function of the form

$$d' = ax^b \qquad (3)$$

where a is a constant and b is the slope of the psychometric function when $\log d'$ is plotted against $\log x$. They found satisfactory fits to all psychometric functions using either $x = \Delta p/p$, $x = \Delta I/I$, or $x = \Delta L$. What was noticeable, however, was that the slope constant, b, appeared to decrease consistently as the threshold value for x increased when $\Delta p/p$ or $\Delta I/I$ was used as the abscissa. Such a decrease was not apparent when $x = \Delta L$; in fact, the slope constant, b, was nearly unity for all experimental conditions. Thus, on the basis of parsimony, one might want to use ΔL as the stimulus measure, since the only parameter of the psychometric function that changes over these stimulus conditions is the scale constant, a, which can be interpreted as the threshold value (the value of x when $d' = 1$).

This is an interesting argument for two reasons. First, it suggests that the slope of the psychometric function is independent of stimulus conditions when ΔL is used to measure the stimulus. Second, it suggests that the slope constant, b, changes when, for example, $\Delta p/p$ is used as the abscissa of the psychometric function. This is contrary to the theoretical expectations of either the optimum detection theory or Laming's theory.

2.4 Weber's Law

In 1834, the German physiologist E.H. Weber announced the generalization that a just noticeable change in stimulus magnitude is a constant proportion of the initial magnitude. Thus, to detect a just noticeable increase in light intensity might require a 2% increase in level. The same summary is true for a number of other sensory continua. It seemed such a solid generalization that Fechner (1860) made it the central assumption of his theoretical structure. In audition, Weber's Law does not provide a very good summary for the premier auditory stimulus—the sinuosid. For intensity discrimination of a single sinuosid, Weber's Law is only a first-order approximation, since the supposedly constant fraction decreases fairly consistently as the base intensity of the stimulus increases. This auditory anomaly is commonly referred to as the "near-miss to Weber's Law" (McGill and Goldberg 1968). The simplest approximation to the bulk of the data is of the form (Green 1988, p. 56)

$$\Delta p/p = \tfrac{1}{4}(p/p_0)^{(-1/6)} \qquad (4)$$

where Δp is the just-detectable increase, p is the base pressure of the sinu-

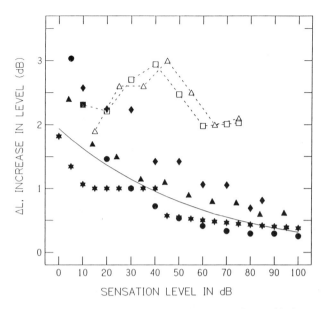

FIGURE 2.2. Data from several experiments on the just detectable increment in the intensity of a sinusoidal signal. The sensation level of the signal is given on the abscissa. The threshold for the increment is measured as ΔL (see text). The experiments are as follows: Riesz (1928)—filled circles; Rabinowitz et al. (1976) approximation—filled stars; Florentine (1983)—1000 Hz, filled triangles; 14,000 Hz, open triangles; Florentine, Buus, and Mason (1987)—1000 Hz, filled diamonds; 14,000 Hz, open squares; Jesteadt, Weir, and Green (1977) approximation—solid line.

soidal stimulus, and p_0 is the absolute threshold of the sinusoid. When we compute $20 \log(p/p_0)$, we are computing the *sensation level* of the pressure p. The approximation expressed in Equation 4 is reasonably accurate for sinusoidal frequencies between 200 and 8000 Hz and for a signal duration of about 500 ms. At or near threshold ($p/p_0 = 1$ or the sensation level of 0 dB), $\Delta p/p$ is about 0.25 and, at the 100-dB sensation level ($p/p_0 = 10,000$), $\Delta p/p = 0.0538$, a change of a factor of five, so Weber's Law is only approximately correct for this intensity change.

Figure 2.2 shows how ΔL changes with the sensation level in dB for several empirical investigations or approximations to the data suggested by different authors. The approximation of Equation 4 is shown as a solid line in Figure 2.2. Riesz's (1928) data at 1000 Hz are shown (filled circles), and the Rabinowitz et al. (1976) approximation to many studies is also illustrated (filled stars). Florentine's (1983) data at 1000 Hz are also shown (filled triangles) and fall reasonably near the other points; however, the data at 14,000 Hz are quite different (open triangles; see Section 2.4.2). Florentine, Buus, and Mason (1987) also measured the Weber fraction for tones of 10 different frequencies (250 to 16,000 Hz) and a large range of sound pressure levels. Some of their more recent data are shown in Figure 3.2 (filled diamonds—1000 Hz, open diamonds—14,000 Hz). While there is reasonable agreement among these studies/approximations, the average data of the studies clearly show considerable scatter. Rabinowitz et al. (1976) normalized the data of 15 studies to obtain an estimate of the threshold value at a 40-dB sensation level. The average value of $\Delta p/p$ is about 0.15, but the range is from 0.1 to 1. Green (1988) attempted to further correct the estimates using a duration correction but only managed to reduce the range from 0.1 to 0.3. A number of stimulus variables that might be responsible for such a large range of estimates were considered, but it was concluded that differences among listeners were the most probable source of the discrepancies.

Differences in these estimates can, of course, be minimized by using measures other than $\Delta p/p$. One of the more compressive is ΔL. Zwicker and Henning (1985) recently published data on 10 different listeners detecting an intensity increment in a 250-Hz sinusoid at several different intensity levels. While the average listener had a threshold of about 1.0 dB, the range from the most sensitive listener to the least was from about 0.5 to 2.0 dB in ΔL. This corresponds to a difference in $\Delta p/p$ of about 0.05 to 0.25 or, if $20 \log \Delta p/p$ is used as the metric, a change of 7 dB. Thus, the reader should be aware that statements about the consistency of Weber fractions, or the lack thereof, are highly dependent on the metric used to measure the difference limen.

As Figure 2.2 suggests, the data obtained with the continuous presentation method (solid circles) may be slightly different from those obtained with the gated presentation method used in more recent studies. The continuous thresholds are larger at the lowest sensation levels and produce a somewhat flatter function at the high-intensity level than do data using a pulsed presentation method. There is little that can be done to reduce variability among different estimates, but stimulus variables, other than base intensity, known to affect the size of the difference limen can be summarized.

2.4.1 Duration

Riesz used a procedure where the stimulus was constantly present, so it was difficult to determine the effective duration of the stimulus. With modern techniques, the listener hears two sinusoids presented for a fixed duration, T. How does the difference limen for intensity depend on this duration? There are two older studies by Garner and Miller (1944) and Henning (1970) as well as the more recent, and extensive, study by Florentine (1986). All agree that the ability to hear small changes in intensity improves as the duration of the stimulus increases. Florentine's data can be summarized in the following manner: ΔL decreases with a shallow slope (about -0.25 versus $\log T$) until a duration of 1 s or more is reached, then ΔL is constant or increases very little with further changes in duration. Florentine found this general relation to hold for frequencies of 250, 1000, and 8000 Hz and for sound-pressure levels of 40, 65, and 85 dB. This is a very long integration time as compared with data obtained at absolute threshold (Plomp and Bouman 1959; Zwicker and Wright 1963; Watson and Gengel 1969). Those data indicate that, for a constant intensity of signal, the thresholds decreased as signal duration increased to about 200 ms where they appeared to be asymptotic. There is some dispute about whether the function is the same for all frequencies (for a summary see Watson and Gengel 1969). The same relatively short time constant (100 to 200 ms) is apparent in detecting sinusoidal signals partially masked by noise (Green, Birdsall, and Tanner 1957). Berliner, Durlach, and Braida (1977) also measured improvement in the difference limen as signal duration changed from 0.5 to 1.4 s. Thus, it appears that the ear has a relatively long integration time for an intensity increment in a sinusoidal signal.

2.4.2 Frequency

It has been known since Riesz's original experiment that the decrease in the difference limen as a function of intensity was very similar at all frequencies if the sensation level (pressure level re: threshold at that frequency) was used as the measure of base intensity. Signal frequency produced no statistically significant differences in the data of Jesteadt, Wier, and Green (1977), and they suggested that Equation 4 held for all signal frequencies used in their study (200 to 8000 Hz). In 1972, Viemeister proposed a theory to explain why the difference limen decreased as base intensity was raised. He suggested that, as base intensity increased, higher-order distortion products became audible. Changes in the intensity of the sinusoid produced larger relative changes in the intensity of these distortion products and, hence, ΔI should decrease as I is increased. Motivated by this suggestion, several investigators have measured difference limens for tones having frequencies in excess of 10,000 Hz, where any quadratic or cubic distortion products will occur above the range of human hearing. Since these distortion components are inaudible, they cannot influence the function depicted in Figure 2.2. Unfortunately, the results of this apparently simple test have been contradictory.

Schacknow and Raab (1973) found the same decrease in the difference limen with increases in base intensity at 250, 1000, 4000, and 7000 Hz. The lower frequencies might be affected by distortion products, but the higher frequencies could not be. Penner et al. (1974) found essentially the same pattern of results; the difference limen decreased with increases in base intensity at 150, 250, 1000, 6000, 9000, and 12000 Hz. Thus, the results of these two studies are parallel to the solid line of Figure 2.2 for all frequencies and inconsistent with Viemeister's hypothesis. A more recent study by Florentine (1983), however, produced data supporting Viemeister's distortion theory. Her data are displayed in Figure 2.2. At 1000 Hz, her data (open circles) are typical of a number of other studies, as summarized by the solid line. At

14000 Hz, the data (shown as open triangles connected by a dashed line) are quite different. The difference limen is larger at the higher frequency and the thresholds do not decrease as base intensity increases. Long and Cullen (1985) also report anomalous results for high-frequency sinusoids similar to those found by Florentine. Other than the differences in listeners, there are no major differences in procedures among these studies.

While such discrepancies remain a hallmark of studies on the difference limens for sinusoidal signals, the data obtained when noise is used as the masker are a model of consistency. These studies are discussed in Section 3. But first, a brief comment on the the use of intensity discrimination data to calculate the "loudness" of the sound.

2.5 Loudness and Intensity Limens

Fechner (1860) was the first to suggest that if a sound's loudness grows as the logarithm of intensity, then equal steps in loudness (just detectable increments) would correspond to equal ratios of intensity and $\Delta I/I$ would be constant, which is Weber's Law. The assumption that $L = k \log(I)$, where k is a constant, L is loudness, and I is intensity, is called Fechner's Law. In the intervening years, this "law" has been widely discussed, both mathematically (Luce and Edwards 1958) and scientifically (Boring 1950). The bulk of our present data on direct estimates of sound loudness (Stevens 1975) suggests that sound loudness grows according to a power function, $L = c I^r$, where c is a constant, L is loudness, I is intensity, and the exponent, r, is 0.3. If stated in terms of sound pressure, p, the power function becomes $L = c p^{2r}$.

In audition, not only did McGill and Goldberg (1968) coin the term "near miss to Weber's Law," they also suggested that intensity discrimination data might be interpreted theoretically to calculate a sound's loudness. Basically, they assumed that the discrimination of a change in intensity was determined by a statistically reliable change in the neural count, which they assumed was a Poisson process. The total number of neural counts, n, generated by a particular sound could be interpreted as the "loudness" of the sound. Some straightforward calculations led to the following interesting conclusion. If it is assumed that Weber's Law is only approximately true, as Equation 4 suggests, then the exponent of that equation (1/6) should be equal to the exponent, $r = 0.3$, of Stevens' power law. McGill and Goldberg noted that these quantities were not equal but observed considerable variability in the estimates of these constants. Hellman and Hellman (1986) have discussed these discrepancies and the approximations used in the derivation of these relations. For more information, consult the recent paper by Hellman and Hellman (1990) that provides an extended analysis of these ideas. Also, the paper by Zwislocki and Jordan (1986) should be considered because it suggests that the slope of the loudness is independent of the slope of the intensity discrimination function. While no simple summary is available as yet, the century-old thoughts of Gustav Fechner are clearly alive and well (see also Moore, Chapter 3).

3. Intensity Discrimination Tasks with Noise Maskers
3.1 Introduction

Noise is the great leveler. Different studies of the thresholds for pure tones partially masked by noise often agree within one decibel and show essentially no difference among listeners. The consistency of the data as well as a highly developed theory of such detection tasks have generated considerable interest in this topic over the past several decades. Our discussion will be restricted to maskers that occupy an extended frequency range and are essentially flat. Filtered noise and pure tones have been widely used as maskers to study various aspects of auditory frequency analysis and will be discussed in Moore's chapter (see Chapter 3).

Unlike data on the difference limen for sinusoidal stimuli, the topic of noise masking has always had a strong theoretical component. The development of radar during World War II stimulated the formulation of a quantitative theory of signal detectability (Middleton 1954; Peterson, Birdsall, and Fox 1954). The theory derived the relation between physical parameters of the signal and noise and the detectability of such signals. It had a strong influence on the techniques used in psychophysical investigations, both on data collection and analysis (Green and Swets 1966). These same techniques helped to refine a theoretical construct first suggested by Fletcher (1940), what he called the "critical bands." These critical bands are the first stage of filtering in the auditory process and would naturally exert a profound influence on any number of psychoacoustic phenomena. Zwicker (1952, 1953) explored the implication of this filtering in such diverse areas as phase-perception and loudness summation. Good summaries of these efforts are found in Zwicker, Flottorp, and Stevens (1957) and Scharf (1970). Both the theoretical analyses and the empirical data suggest that a fundamental quantity of importance in a noise masker is N_o, the spectral level of the noise.

3.2 Noise Power Density, N_o

While there is some diversity in our specification of the "Weber fraction" for pure tone stimuli, there is essentially no disagreement as to the important noise parameter. The critical quantity is the noise power density, or spectrum level, of the noise. If the noise has a flat power spectrum over a band 0 to W Hz, then the total power of the noise, N, is

$$N = N_o W \qquad (5)$$

The bandwidth, W, has dimensions of Hertz (cycles per second, T^{-1}); thus, N/W is a power multiplied by time and N_o has the dimension of acoustic *energy*. It is, of course, the average noise power that one would measure in a frequency band 1-Hz wide. Because the noise waveform is statistical in nature, it is best estimated by measuring over a large bandwidth, W, for as long an integration time as possible. Then N_o can be estimated by dividing N by W. Before the fundamental significance of this quantity was recognized, it was common to measure only the overall power of the noise, N. Fortunately, the bandwidth of the noise was often cited, and it is possible to calculate N_o from such studies.

3.3 Optimum Signal Detection Formulae

Signal detection theory derived formulae relating the ratio of signal energy to noise power density, E_s/N_o, for a variety of different detection devices. These devices were all optimum detectors because they used likelihood ratios as the decision statistic to decide whether signal-plus-noise or noise-alone was present. Several different detectors were analyzed; they differed in the assumptions made about the degree of knowledge that the detectors possessed concerning the signal. The most extensive knowledge was assumed for the signal-known-exactly case. Here the detector knew all aspects of the signal completely. For a sinusoidal signal, the detector knew the starting-time, duration, frequency, and phase. The optimum detector was a matched filter, or cross-correlation detector, and the resulting likelihood ratio was distributed as two Gaussian distributions separated by a normalized distance, d'. For this signal-known-exactly case, d' is given by

$$d' = \sqrt{\frac{2E_s}{N_o}} \qquad (6)$$

If some of the signal parameters were uncertain, then different detection formulae would result. For example, if the frequency of the signal were unknown, then the detector must correlate with each of the potential signal waveforms and combine the results of these calculations to decide about the presence or absence of the signal. The resulting d' formula would be altered, and the ratio of E_s/N_o must be increased to achieve the same detectability performance (Green and Birdsall 1978). For our purposes, only two other detection formulae need be discussed. One is the detectability index for an energy detector. In this case, the detector is simply monitoring the energy in a band, W, with noise density, N_o, and trying to detect a potential signal of energy, E_s, and duration, T. Pfafflin and Mathews (1962) and Green and Swets (1966) have analyzed this case. To a good first approximation

$$d' = \frac{1}{\sqrt{WT}} \frac{E_s}{N_o} \qquad (7)$$

Note that d' increases with signal energy at a different rate for the two equations. In Equation 7, d' is proportional to signal energy; in Equation 6, d' is proportional to the square root of signal energy, and, because signal energy is proportional to the sound pressure squared, d' is proportional to signal pressure.

Our final detection formula concerns an extreme in signal uncertainty. Suppose the signal is a sample of noise with noise power density, S_o. This signal is added to another, uncorrelated noise, N_o. The only known parameters of the signal are its duration, T, and bandwidth, W, since the waveform is a random process. The detection formula for that case (Green 1960) is approximately

$$d' = \sqrt{WT}\, \frac{S_o}{N_o} \qquad (8)$$

where W is the bandwidth of the signal or noise and T is its duration. The optimum detector for this noise waveform is, in fact, an energy detector. The total energy of the unknown noise signal is $E_s = WTS_o$, so if we substitute for that quantity in Equation 7, we produce Equation 8.

For all these detection cases, the distribution of signal-plus-noise and noise-alone are approximately Gaussian. The variances of the two distributions are the same for the signal-known-exactly case. For the energy detector, the signal-plus-noise variance is larger than the noise-alone variance by an amount that depends on the signal-to-noise ratio. If that ratio is small, as it often is, then the variances are nearly equal.

3.4 Sinusoid Signal in Noise

The detectability of a sinusoidal signal in noise has occupied an important place in early psychoacoustic studies. Fletcher (1940) used such a signal to estimate the width of the critical band is his classic study. Hawkins and Stevens (1950) used this signal and measured its detectability over a wide frequency range and at several different overall levels of noise. The earlier studies did not control the signal duration. They simply presented the sinusoid continuously and had the listener adjust the tone until it was just audible (Fletcher) or had a "just noticeable pitch" (Hawkins and Stevens).

3.4.1 Duration

The detectability of a sinusoidal signal in noise depends on the duration of the signal. If signal energy is held constant, then detectability is constant for duration between about 10 and 100 ms (Green, Birdsall, and Tanner 1957). At shorter durations ($T < 10$ ms), there is considerable splatter of the energy about the signal frequency, and the signal energy falls in many different critical bands. Thus, at these short durations, signal energy must be increased to hold detectability constant. At longer durations, ($T > 100$ ms), the human detector apparently cannot integrate all the signal power, and signal energy must again be increased to hold the detectability constant. Thus, at very long durations, the detectability depends only on signal power.

3.4.2 Frequency

Detectability of a sinusoidal signal depends on the frequency of the signal. Lower frequency signals are easier to hear than higher frequency signals, presumably because the critical band widths are smaller at the lower frequencies. Empirically, we know that to hold $d' = 1$ as the frequency is changed requires the following change in E_s/N_o

$$10 \log E_s/N_o = 8 + 2 \, (f/F) \tag{9}$$

where f is the signal frequency in Hz and $F = 1000$ Hz, at least for signal frequencies between 250 and 4000 Hz (Green, Licklider, and McKey 1959). They used a 100-ms signal in their study and their signal power at $d' = 1$ is about the same as the threshold power obtained by Hawkins and Stevens (1950) and by Fletcher (1953) who used continuous signals.

3.4.3 Multiple Sinusoidal Signals

Suppose the signal to be detected is a complex signal consisting of two or more sinusoids. How does the detectability of the complex depend on the number and detectability of the individual components that comprise the complex? If the individual sinusoids all fall within the same critical band, then the detectability of the complex is determined simply by the total energy of the signal (Gassler 1954; Marill 1956). Thus, for two sinusoids of equal detectability, the energy of each can be decreased by 3 dB, and the detectability of the complex remains unchanged. All investigators agree on this result.

What was less certain was the detectability of components that fell in different critical bands. Gassler (1954) and Marill (1956) claimed that the detectability of two sinusoidal signals was no better than the more detectable sinusoid composing the complex. Green (1958) claimed that a kind of statistical integration occurred if the signal fell in different critical bands and that the combination was about 1.5 dB lower if both were equal in detectability. Green, Licklider, and McKey (1959) combined as many as 16 sinusoids and found that the complex was definitely more audible than any single component. They suggested that the combined detectability is equal to

$$d'_c = \sqrt{\sum_{i=1}^{m} d_i'^2} \tag{10}$$

where d'_c is the detectability of the complex, d'_i is the detectability of the individual sinusoids, and m is the total number of such sinusoids. Note that

d'_c is equivalent to the magnitude of a resultant vector formed from vector addition of the individual, orthogonal vectors. One rationale for this formula is that each component is independently detected, and the decision about the complex is based on some combination of the individual decisions (post-detection integration).

With 16 equally detectable components, the energy of each can be decreased by 6 dB [5 log(m)], and the detectability of the complex remains unchanged. Since the total energy of 16 equal-amplitude components is 12 dB greater than the individual energy, it is clearly wasteful to distribute signal energy into different spectral regions. Detection is best when the signal energy is concentrated in a single critical band.

Later research has appeared to confirm the vector combination rule for the detection of multiple components in noise. Buus et al. (1986) studied the detection of an 18-tone complex in a uniform masking noise. They found the complex tone about 6 dB more detectable than single sinusoidal components. In a recent study of the detection of brief sinusoidal signals shaped by a Gaussian envelope, van den Brink and Houtgast (1990) found that long duration (≈ 100 ms) pulses followed the 5 log(m) rule suggested by Equation 10. Interestingly, they found that for much shorter duration signals (< 10 ms), the data followed an even stronger integration rule, 8 log(m).

3.4.4 Psychometric Functions

Historically, the function relating the proportion of correct detection decisions to some measure of signal size or magnitude is known as the psychometric function. This relationship will now be considered for various signals added to noise and for an increment in the amplitude of the standard signal, the differential intensity limen.

The detectability index, d', can be related to the proportion of correct decisions in simple detection tasks. For example, if a two-alternative forced-choice procedure is used, one drawing comes from the signal distribution, call it s; the other comes from the noise-alone distribution, call it n. The probability of a correct response is the probability that $s > n$, or $s - n > 0$. If s and n are Gaussian, the difference distribution is also Gaussian, and the proportion of correct responses, $P_2(c)$, is

$$P_2(c) = \Phi\left(\frac{d'}{\sqrt{2}}\right) \tag{11}$$

where Φ is the Gaussian integral (the $\sqrt{2}$ arises because the variance of the difference distribution is twice the variance of the original distributions). Thus, for $d' = 1$, the expected proportion of correct responses is about 0.76. There are several tables relating d' to the percentage of correct responses in two-alternative forced-choice tasks (Green and Dai 1991). Because the relation between d' and E_s/N_o is known for several detectors, it is natural to compare the human listeners' performance with that of various detectors.

3.4.4.1 Sinusoid in Noise

There is a great deal of data showing how the percentage of correct detections varies with signal energy for a single sinusoid added to wideband noise. To a good first approximation, d' increases as E^k, where k is a constant. In the case of a sinusoidal signal masked by clearly audible noise, k is approximately unity, a fact that has been used to argue that the system is an energy detector (see Eq. 7). A typical psychometric function of a sinusoid masked by noise is shown in Figure 1 as the solid circle points. The dotted line is generated from Equation 10, where d' is determined from the energy detector, Equation 6. When plotted as a function of $10 \log(E_s/N_o)$, the psychometric function rises from 60% to 95% in about 6 dB, or a slope of roughly 6%/dB. In Green, Licklider, and McKey (1959), there are many examples of such psychometric functions for different individuals and signal frequencies.

For a sinusoid presented in quiet, the psychometric function is apparently much steeper. The value of k depends on signal frequency but is in the range of 1 to 3 (Watson, Franks, and Hood 1972). The trading ratio is therefore about 6%/dB at the lower frequencies ($f < 500$ Hz), but as small as 4%/dB at the higher frequencies ($f < 1000$ Hz).

3.4.4.2 Noise Signal in Noise

Green (1960) has argued that a steeper psychometric function may occur because the listener is uncertain about some parameter(s) of the signal. If the optimum detector is uncertain about some signal parameter, the signal energy must be increased to produce a given value of d', and the slope of the psychometric function also increases (Green 1960). Green and Sewall (1962) studied the psychometric functions produced when the signal was a noise sample. Since the bandwidth of the noise signal was the same as the background noise, the only signal parameter that could possibly be uncertain was the starting time and duration of the noise. When the signal was added to a continuous masking noise, the psychometric functions were slightly steeper than Equation 7 predicts. If a "gated" condition was employed, in which the signal and noise were gated on and off together, there could no longer be any temporal uncertainty. The listener heard two noise bursts and simply tried to pick the more intense burst. In that situation, the slopes of the psychometric functions were the same as those predicted by Equation 8; d' increased proportionally with S_o/N_o or signal energy.

3.4.4.3 Increment in a Sinusoidal Signal

For the sinusoidal signal, one method of removing uncertainty is to add the signal to another sinusoidal signal (often called the pedestal). Because the pedestal is audible in the background noise, the signal frequency is known. The pedestal is gated to remove any temporal uncertainty. Thus, the listener hears two sinusoidal bursts and must select the more intense because the signal, when added inphase to the pedestal, will increase its energy or power.

The small cartoon in Figure 1 illustrates this procedure. The psychometric function obtained is illustrated by the solid square points in Figure 2.1. It is virtually the same as that expected from the optimum (cross-correlation) detector (see Eq. 7; Green and Swets 1966). The psychometric function rises at about 2%/dB when the abscissa is plotted in terms of $10 \log E_s/N_o$. The similarity of the data to the expected psychometric function does not mean that the listener is an optimum (cross-correlation) detector. This same psychometric function can also be derived by assuming that the listener is an energy detector. In effect, the contrived listening conditions make energy detection equivalent to cross-correlation detection. This same psychometric function occurs if there is no noise background and the listener is simply trying to determine which of two tones is more intense—difference limen experiment (Hanna, von Gierke, and Green 1986). Again, the psychometric function is a very shallow 2%/dB compared with the 6%/dB obtained for a sinusoid in noise. Laming (1986) has interpreted this consistent change in the slopes of the psychometric function to reflect what he calls differential coupling. His central thesis is that a "change" detector plays a very central role in all sensory phenomena.

4. Complex Signals

4.1 Introduction

The signals used in the first half-century of quantitative research on the topic of auditory intensity discrimination were simple—largely single sinusoids or noise. The material covered in the first three sections of this chapter is a summary of those research findings. The items of primary concern were the threshold values, Weber's Law, and psychometric functions. The intensity discrimination task involved essentially a *successive* comparison of two auditory sounds. During the past decade, there has been a considerable shift in the emphasis of research on auditory intensity discrimination. In this author's view, the change in emphasis can be traced to a basic change in the investigators realizing the importance of simultaneous comparisons in the process of intensity discrimination. This view led to an investigation of discrimination tasks requiring a *simultaneous* comparison of different spectral regions, rather than comparing the intensity of successively presented sounds. In principle, the comparison process could involve the presentation of a single sound. The listener's task would be to compare the intensity levels at two different spectral regions. Because the sound is more complicated, consisting of energy at two or more spectral locations, the stimulus is often described as complex. Thus, the phrase "complex auditory signals" has been used as a general label for this new research emphasis. Four distinct areas of research can be identified. These are: (1) spectral shape discrimination (profile analysis); (2) comodulation masking release, CMR; (3) comodulation detection difference, CDD; and (4) modulation detection interference, MDI.

Green (1983) and his colleagues have studied the discrimination of a change in spectral shape—what they call "profile analysis." A typical task involves a multicomponent complex of 10 to 20 sinusoidal components widely separated in frequency. For the standard stimulus, all components of the complex are equal in amplitude. For the standard-plus-signal stimulus, a sinusoid might be added, inphase, to a single component, thus increasing the intensity of one component of the complex. To detect the difference between the standard and standard-plus-signal spectra, the listener at the very least must be able to simultaneously compare the intensity level at two spectral locations. Interestingly, many listeners can hear a change in spectral shape better than they can hear a change in level of a single sinusoidal component, that is, the threshold in the profile task is typically smaller than the Weber fraction (see also Yost and Sheft, Chapter 6).

Hall, Haggard, and Fernandes (1984) used the phrase comodulation detection to describe their experimental procedure. They measured the listener's ability to hear a sinusoidal signal added to a narrow band of noise, much as had been done in Fletcher's classic study of the critical band (Fletcher, 1940). The threshold for such a signal depended strongly on the relationship between the envelope of that narrow band and the envelope of other bands located at different frequency regions throughout the spectrum. When an adjacent band was modulated by the same envelope as the signal band, then the threshold of the signal was about 10 dB lower than if the adjacent band had been modulated by an unrelated envelope. The presence of an adjacent, comodulated band of noise made the signal considerably easier to hear. Obviously, such improvement depends on the ability of the observer to simultaneously compare the envelope at two spectral locations, the signal band and the adjacent band. Thus, in this comodulation task, the listener is apparently simultaneously comparing the intensity fluctuation in the level of at least two different spectral regions. Comodulation detection could be described as a dynamic form of profile analysis.

A curious relative of CMR is found if the detection of one noise band is measured in the presence of a second noise band. Cohen and Schubert (1987b) and McFadden (1987) were the first to systematically explore this area. They found that the signal band was easier to detect when the envelopes of the two bands were unrelated rather than when they were correlated. The differences in thresholds are called comodulation detection differences (CDDs), and it should be noted that these results are in some sense the opposite of what is found in CMR, since coherence of the envelope in the two frequency regions makes detection of one more difficult.

In the final experimental paradigm, Yost and Sheft (1989) have shown that detection of amplitude modulation of a carrier (target) tone is strongly affected by the presence of modulation at another (flanking) tone, even if the flanking tone is widely separated in frequency from the target tone. Since modulating the flanker interferes with the detection of modulation at the target tone, they call this modulation detection interference (MDI). The review of MDI will be very brief because it is discussed at length by Wightman and Kistler (Chapter 5) and Yost and Sheft (Chapter 6).

In Sections, the major experimental findings from each of these areas will be briefly reviewed. As has been seen from their descriptions, they may all involve a simultaneous comparison of the intensity level at two or more spectral regions.

4.2 Spectral Shape Discrimination (Profile Analysis)

4.2.1 Successive Versus Simultaneous Comparison

One of the earliest profile-analysis experiments nicely illustrates the fundamental difference, both in procedures and results, between the traditional intensity discrimination paradigm and the new experimental paradigm (Green, Kidd, and Picardi 1983). Let us begin with the results of the older, traditional experiment on intensity discrimination. There are stimuli: a) a 1000-Hz sinusoid, and b) a 21-component complex. The 21 components are all equal in amplitude and equally spaced in logarithmic frequency between 300 and 3000 Hz. The duration of the sounds is 100 ms. The ability of the listeners to hear an increase in intensity (Weber fraction) for either of these standards is measured by presenting the standard in one interval and the standard-plus-increment in the other interval of a two-alternative forced-choice task.

The new experimental paradigm departs somewhat from the traditional one in that the standard level is chosen from a 40-dB range on each trial (roving level). The independent variable of the experiment is the time between these two presentations, the interstimulus interval. It ranges from 100 ms to 8000 ms. Figure 2.3 displays the results of this experiment. The upper two curves show the results for the 21-component complex (solid circles) and the 1000-Hz tone (open circles). The abscissa is the interstimulus interval in milliseconds. The ordinate is the threshold for discriminating a change in intensity. It is measured as the signal-to-standard ratio in decibels $[20\log(\Delta p/p)]$. In the explorations of any new area, it is prudent to be sensitive to possible differences among conditions, rather than to assume everything is much the same. Thus, $[20\log(\Delta p/p)]$ has been consistently used as the dependent variable in the investigations of profile analysis. It is one of the least compressive measures of the Weber fraction. The standard error of this measure is also nearly constant (about 1 to 3 dB for a typical measurement) and independent of the threshold estimate, at least for the range of -30 to 0 dB. Obviously, later empirical results or theories might convince us of the wisdom of using some other metric.

Clearly, the interstimulus interval influences the threshold value, as previous researchers have demonstrated. The best threshold value is somewhat higher than is typically measured in a traditional intensity discrimination experiment. This increase is caused by the 40-dB range of the standard level. The threshold value is typical of that observed by Berliner and Durlach

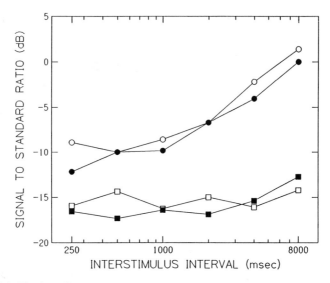

FIGURE 2.3. The just discriminable increase in intensity for a variety of conditions as a function of the interstimulus interval, the time between the two sound presentations in a two-alternative forced-choice task. The ordinate is the signal threshold measured as the signal-to-standard ratio (see text). The filled circles show results for a single 1000-Hz tone with a roving level between trials. The open circles show the same results except there are 21 components in the complex. The open square is a profile task, an increment in a single tone of a 21-component complex. There is random variation of intensity on each presentation. The solid square is the same condition, except the intensity variation occurs between trials.

(1973) for this range of roving levels. At the longer interstimulus intervals, the increase in threshold level presumably reflects the decay in the precision of the memory trace for the first presentation level. These pure intensity discrimination results can be contrasted with those obtained in two other conditions. The stimuli are similar, but the discrimination is not only a change in intensity level but a change in spectral shape as well.

In these two conditions, the standard is always the 21-component complex. The signal is always an increase in the level of the 1000-Hz component, thus producing a bump in an otherwise flat spectrum. The independent variable is again the interstimulus interval. In one condition, the level is always chosen from a 40-dB range for each presentation (within-trial variation). In the second condition, the level is changed between trials. When the level variation is between trials, the listeners could be comparing the level at 1000 Hz in the two intervals, just as they had in the previous experiment. The results for this condition are shown as the solid squares (Fig. 3). For within-trial variation of level, the results are shown as the closed squares. To achieve such detection accuracy, the listener must discriminate spectral shape independent of the overall level of the stimuli. Two important points are evident.

First, the threshold values are lower than for either of the pure intensity discrimination tasks (circle points Fig. 3). It is easier to hear a change in spectral shape than an increase in the intensity of a single sinusoid. Second, the threshold is nearly independent of the interstimulus interval.

Clearly, the listener is not estimating intensity level as in the first two experiments but is estimating a change in spectral shape—a change in the quality of the sound. Some people have called it timbre perception. This qualitative change is largely independent of the overall level of the sound. Just as an oboe or flute retains its qualitative character despite changes in overall level, so detection of a change in spectral shape (profile analysis) seems to be largely impervious to changes in overall level.

The change in spectral shape is not like the estimate of intensity level required in traditional experiments. The change in spectral shape produces a qualitative difference in the sound, which apparently can be coded in a categorical manner. Because the threshold with an interstimulus delay of eight seconds is nearly the same as the threshold obtained with an interstimulus delay of one-quarter second, it may be inferred that the memory of this auditory quality does not appear to deteriorate appreciably with time.

4.2.2 Phase and Frequency Effects

Given that the listener is reasonably sensitive to a change in the intensity of a single component of a multitonal complex, is the basis for such a discrimination simply the change in spectral level? An alternative form of this question is whether the waveform of the complex is important, that is, whether the phase relationship among the components influences the detection of a spectral change. In some experiments designed to directly test this question, Green and Mason (1985) compared the detection of an increment in the central component of a multitonal complex in two phase conditions. In the first condition, the phases of all components of the complex were fixed—this is the procedure used in all the earlier experiments on profile analysis. When the phase is fixed, the standard stimulus always has the same wave shape. The addition of a signal to the standard alters that wave shape. In the second condition, the phases of each component of the complex were randomly selected on each presentation. The waveform was completely unpredictable; only the difference in the power spectrum distinguishes the standard from standard-plus-signal. For the standard stimulus, the components were all equal in amplitude. For the standard-plus-signal stimulus, the middle component of the complex was increased in level. The results showed that the thresholds for these two conditions were virtually the same; thus, phase is irrelevant. The relevant stimulus for this discrimination task is the power spectrum. As better and faster computers have become available, it has become standard practice to randomize the phase relation among the components. So far as anyone knows, it is irrelevant to the discrimination of a change in spectral shape.

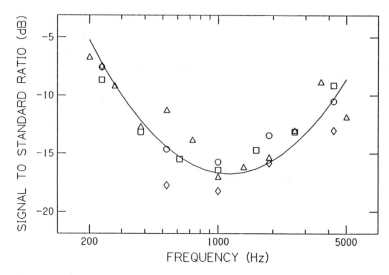

FIGURE 2.4. The results of several experiments in which the signal is an increment in a multitonal complex. The threshold value is measured as the signal-to-standard ratio. The frequency of the single increment is given on the abscissa. The symbols are data points measured in different experiments Green and Mason (1985)—triangles; Green, Onsan and Forrest (1987)—squares; Bernstein and Green (1988)—diamonds; Bernstein and Green (1987)—circles.

A second question, also first investigated by Green and Mason (1985), concerns signal frequency. Most of the previous experiments had created a spectral change by increasing the central component of a flat (equal-amplitude) multitonal complex. What if the spectral bump is moved to other regions of the spectrum? Although there is variability among the different listeners, it is generally true that a change in the central component of a multitonal complex is easiest to detect. If one plots the threshold for a spectral bump in an otherwise flat spectrum, a bowl-like function is evident. Figure 2.4 shows the data taken from four experiments. In all these experiments, there were 21 equal-amplitude components ranging over a large frequency range—200 to 5000 Hz is typical. An adaptive procedure was generally used, and the threshold value is given at a percent correct value of about 70%. A bowl-like function is apparent in the results. The solid line is a simple approximation suggested by Bernstein and Green (1988). It is of the form

$$20 \log \Delta p/p = 20 [\log(f/1150)]^2 - 17.50 \qquad (12)$$

Note, the square operates on the logarithm of the frequency ratio. Again, this function is typical of data averaged over a number of listeners. Each listener is likely to show a distinctive function that is characteristic of that listener and only similar to the average.

A number of experimental variables have been studied in an attempt to understand this bowl-like function. Both Green and Mason (1985) and Green, Onsan, and Forrest (1987) suggested that the absolute frequency of the signal is not the sole variable responsible for the bowl-like function. They changed the frequency range of the entire complex and showed that the resulting thresholds were also influenced by the relative position of the signal frequency with respect to the range of frequency components in the complex. For moderate frequencies, it appeared that spectral changes in the center of the complex were more easily detected than changes at either extreme. Bernstein and Green (1988) have shown that the bowl appears for complexes containing different numbers (3 to 21) of components. There is still no good understanding of why this function occurs. As we will discuss later in the text, the function may arise because of the weight assigned by a listener to the intensity levels measured at the various components of the complex. It may simply reflect a predilection on the part of the listener to favor some components more than others.

4.2.3 Theory and Spectral Weights

There are two major theories of how the listener discriminates changes in the spectrum. One is Feth's (1974) envelope weighted average instantaneous frequency (EWAIF) theory. It assumes that the listener computes the pitch of a complex sound by weighting the instantaneous frequency by the amplitude of the envelope (see Green 1988, pp. 125–127). In certain circumstances, adding a signal to the standard spectrum alters the pitch. Thus, the theory claims that this pitch change is the basis of the listener's discrimination, even if the spectral change is produced by a change in the intensity of one or more components. When the components are all close together, for example, within a critical band, it appears that this theory has some validity. The listeners report that the two spectra have slightly different pitches. Randomizing the rate of the digital-to-analog (DA) converters, which randomizes the frequency of the components, disturbs their discrimination of a spectral change (Berg 1992). But, for profile experiments in which the components are widely separated in frequency, it appears that this theory is less useful (Richards 1987). For widely separated components, which make up the bulk of the profile analysis literature, a channel theory seems more appropriate.

The channel theory was first proposed by Durlach, Braida, and Ito (1986). It describes the process of discriminating between the two spectra as a statistical decision problem. The decision is based on the measured intensity level of the components of the complex stimuli. Do the component levels more closely resemble the standard spectrum or the standard-plus-signal spectrum? This is a standard signal detection problem, with the component levels distributed as multivariate Gaussian variables. The theory makes only a minimal set of assumptions, but derives the optimum decision rule and the optimal detection performance, d', for this decision task. Green (1988, pp. 113–125) provides a simple presentation of the basic assumptions and derives the major results. Only one feature of the theory, the optimal decision rule, is stressed in this discussion.

Consider the profile task where the two alternatives are a standard spectrum with all components equal in amplitude and a signal-plus-standard spectrum with the central component elevated in amplitude. How should we decide between these two alternatives, given only the estimates of level obtained on each trial? Recall that the overall level on each presentation is also a random variable. Somehow the measured levels must be treated in a way to overcome this variation in overall level. An optimum procedure is to determine the average of the levels of the nontarget components and to subtract that average from the level measured at the target (signal) component. Formally,

$$z = L_s - \frac{1}{(m-1)} \sum_{i=1, i \neq s}^{i=m} L_i \qquad (13)$$

where z is the optimum decision statistic, m is the total number of components in the complex, L_i is the estimate of level at the ith component, and L_s is the level at the signal component. The decision variable, z, is monotonic with likelihood ratio; the larger the value of z, the greater the probability that the target component has a larger intensity than the nontarget components. Note that, by computing such a statistic, the decisions are virtually independent of the overall level of the sound.

Berg (1989, 1990) has devised a way to estimate the weight assigned to the elements of a composite stimulus. Berg and Green (1990) have used this technique to estimate the weights associated with the various components of a profile stimulus. A summary of some of their results is presented in Figure 2.5. The figure shows the results obtained with three different listeners (columns) in three experimental conditions (rows). In the first experiment (top panel), the complex consisted of a three-component complex. The frequencies, 200, 1000, and 5000 Hz, are shown along the abscissa, and the ordinates are the estimated weights. In Berg's (1989) procedure, the largest weight is given a value of one, and the other weights are scaled in relation to the largest. For a three-component complex, $m = 3$, Equation 13 says that the optimum weights are unity for the target component and -0.5 $(-1/(m-1))$ for the nontarget components. The optimal weighting pattern is indicated by the starlike symbols in the middle panels of Figure 5. This is almost exactly the weight pattern shown for the second observer. The middle and bottom rows of the figure show the data obtained for a five-component ($m = 5$) and an eleven-component ($m = 11$) complex. Once more, the target component is given the largest weight and the nontarget components are scaled from that value. These optimum values for the nontarget components are again shown by the starlike symbols in the middle panels. Again, the observer's weight pattern closely resembles the optimum weight pattern.

It is possible to quantify how closely the listeners' pattern of weights mimics the optimum weight pattern and thus to estimate the degree to which

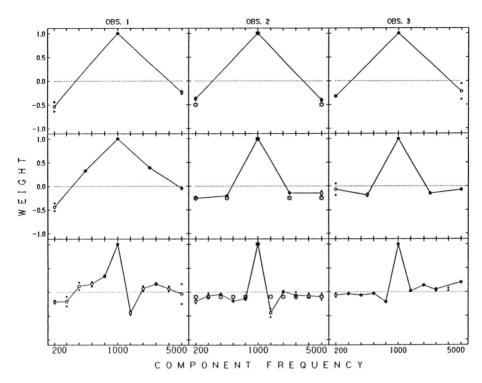

FIGURE 2.5. Spectral weights as estimated in profile analysis experiments. Berg's (1989) COSS technique (see text) is used to estimate the weight given to the level estimate at each component frequency. Those weight values are indicated on the ordinate. The abscissa gives the frequency of the component in the complex. The top panel shows the weight patterns for a three-component complex, the middle panel for a five-component complex, and the bottom panel for an 11-component complex.

the listeners utilize the levels measured at each component in their decisions about the presence or absence of a spectral change. For changes in the intensity of the *central* components of an equal-amplitude complex, the weight patterns tend to be similar to the optimum weight pattern. For changes in the more peripheral components of a multitonal complex, or for more complex changes in the component amplitudes such as ripple or sinusoidal changes, the weight patterns are not as optimal (Berg and Green 1991). Why the listener uses less optimum weight patterns for these other spectral changes is still unclear.

4.3 Comodulation Masking Release (CMR)

In 1984, Hall, Haggard, and Fernandes published a paper that profoundly affected the field of auditory masking. The experimental task was simple: the listener was trying to detect a sinusoidal signal (1000 Hz) added to a narrow band of noise. In one condition, the signal threshold was determined as a function of the bandwidth of the masking noise. The results were as expected: the detectability of the signal was independent of the noise bandwidth when the noise bandwidth exceeded about 130 Hz. For narrower bandwidths, the signal became progressively easier to hear as the bandwidth of the noise was reduced. The results replicated in part Fletcher's classic experiment on the critical bandwidth (Fletcher 1940). In the second condition, these measurements were repeated, except that the band of noise was amplitude-modulated by a narrow (0 to 50 Hz) noise band. Thus, the amplitude of the entire noise masker fluctuated slowly in time, and, what is critical, all parts of the noise spectrum modulated synchronously. For narrow bands of noise, the audibility of the signal was essentially the same as in the first condition. As the bandwidth of the noise increased, however, the signal became easier to hear. More total noise power was causing less masking. At the widest bandwidth, about 1000 Hz, the signal was about 20 dB easier to hear in the synchronously modulated noise than in the unmodulated noise used in the first condition. Synchrony in the spectrum level of the noise produced unmasking. Hall et al. (1984) observed that the primary cue for the detection of the signal appeared to consist of an across-frequency comparison, such as that studied in profile analysis, rather than an analysis of the information at a signal frequency.

In a second experiment, Hall et al. (1984) clearly demonstrated that synchrony of an adjacent band is the critical stimulus ingredient. They fixed the noise band at 100 Hz and centered it at 1000 Hz (this band is usually called the *target* band). In addition to the target band, a second noise band, also 100 Hz wide, was introduced at some other center frequency (this band is usually called the *flanking* band). The critical comparison was the threshold of the signal when the flanking band was an independent noise band and when the flanking band was coherently modulated in amplitude with the signal band. At slow rates of modulation, the signal was 5 to 12 dB easier to hear when the two bands were comodulated compared with the condition in which each band was derived from independent noise sources. This improvement is referred to as comodulation masking release (CMR).

In this review, the stimulus parameters that influence the size of the measured CMR will be examined first. Then, the general class of mechanisms that have been proposed to account for these effects will be reviewed. There is no strong theoretical consensus, and different authors stress different interpretations.

4.3.1 Stimulus Factors—Psychometric Functions and CMR Defined

The psychometric function in CMR experiments was determined only recently in a collaborative experiment carried out both in England and the United States, (Moore et al. 1990). The main purpose of the study was to resolve a persistent difference in the estimated size of the CMR found in the two laboratories. The Cambridge CMR was a bit smaller than the Chapel Hill CMR. Psychometric functions were measured because the two laboratories used different adaptive procedures that tracked different levels on the psychometric function. If the psychometric function for the reference condition (target band alone) was different from the psychometric function determined for the comodulation condition (flanking band present), then the measured CMR would be different. The investigators found that the slopes of the psychometric functions were slightly different: about 3.4% per dB for the reference conditions and about 4.4% per dB for the comodulation condition. Thus, these CMR psychometric functions have a slope somewhat intermediate between the two functions shown in Figure 2.1. The difference, however, was not sufficient to account for the discrepancy in CMR estimates. They concluded that the details of how the noise was constructed (multiplied noise versus a sum of Gaussian components) was the most likely reason for the different CMR estimates.

Before reviewing the stimulus factors, it is important to define what is meant by comodulation masking release. As Schooneveldt and Moore (1987) have noted, there are several possible definitions of the CMR. We can compare the signal threshold measured in the presence of correlated and uncorrelated flanking bands [CMR(U − C) is their notation]. Alternatively, we can measure the threshold of the signal when only the target band is present, sometimes called a reference, R, condition. Then, we might define the CMR as the difference between the reference threshold and the correlated flanking noise threshold [CMR(R − C)]. These will be the same only if the uncorrelated flanking causes no direct masking and the threshold for the signal in the uncorrelated noise is equal to the reference threshold (caveat emptor).

4.3.1.1 Frequency Parameters

Many of the early studies investigated how the frequency difference between the signal band and the flanking band influenced the CMR results. Cohen and Schubert (1987a) used a 1000-Hz signal band with the flanking band located somewhere in the octave above or below the signal band. The CMR was essentially zero when the flanking band was an octave from the signal band but increased as the frequency separation decreased. The CMR effect was slightly larger when the flanking band was below the signal band, and a wider range of flanking frequencies produced a notable CMR. Schooneveldt and Moore (1987) showed a very similar pattern of results for a number of different frequencies for the signal band (250 to 8000 Hz). The largest CMR effects were observed in the 2000 to 4000 Hz region, a result also supported by Cohen and Schubert (1987b).

4.3.1.2 Binaural Presentation

Hall, Haggard, and Harvey (1984), Cohen and Schubert (1987a), and Schooneveldt and Moore (1987) explored a dichotic presentation mode in which the signal band was presented to one ear and the flanking band was presented to the other. The CMR effect was somewhat smaller in this case but clearly present. The frequency range was more broadly tuned for the dichotic than for the monaural case just discussed, but it showed less tuning when the frequencies of the signal and flanking were very close together. In the monaural case, there was direct masking produced by the flanking band on the signal, and the definition of the size of the CMR effect must be carefully considered, as discussed in Section 4.3.1 (see Schooneveldt and Moore 1987). Schooneveldt and Moore (1989) have also investigated a variety of other binaural conditions that cannot be reviewed here.

4.3.1.3 Level and Duration of Signal and Flanking Bands

In most CMR studies, the levels of the signal and masking bands are the same. Cohen and Schubert (1987a) showed that the size of the CMR was generally larger as the noise level increased; CMR was about 10 dB at 75 dB SPL and only 3 dB at 35 dB SPL. Two studies have investigated the relative level of the signal and flanking bands (Hall 1986; McFadden 1986). Both studies found that the largest CMR occurred when the two levels were equal. There is less agreement about CMR's tolerance for differences in level between the two bands. Hall's data showed smaller but measurable CMRs when the difference in level was 30 dB. McFadden found no CMR when the levels differed by 20 dB.

McFadden (1986) showed that the size of CMR remained remarkably constant for signal durations from 75 to 375 msec. In those experiments, the flanking bands were continuously present. He also investigated the masking produced when both flanking and signal bands were gated together. Although the thresholds did not appear to vary as regularly as in the previous experiments, the CMR effect was still evident for durations from 75 to 500 ms. In a final condition, McFadden delayed the envelope of the flanking band with respect to the envelope of the signal band. The CMR diminished sharply after a delay of about 2 to 3 ms. Moore and Schooneveldt (1990) repeated some of these results, but also used much smaller noise bandwidths (6.25, 25, and 100 Hz). They found measurable CMRs only at the smaller bandwidths—the center frequency and band spacing were quite different from those used by McFadden. The CMR decreased as the time delay between the target and cue band increased, but the amount of decrease depended on the bandwidth. The 6.25-Hz bandwidth showed almost no decrease, even at a delay of 20 ms. They correctly observed that the autocorrelation function for the envelope of a narrowband noise process is nearly a monotonic decreasing function of delay and shows none of the periodicity associated with the autocorrelation function of the raw wave-form.

4.3.1.4 Number and Type of Signal Band

Hall, Grose, and Haggard (1988) investigated the CMR for a three-band masking complex in which the signal could occur in one, two, or all three bands. The largest CMR occurred for a single signal presented in a three-band masker, but a measurable masking release could be obtained when three signals were presented in all three bands. Hall and Grose (1988) demonstrated a clear CMR for a signal which consisted of increasing the level of the noise in the target band. Thus, a sinusoidal signal is not necessary for a comodulation masking release. This result is important because, as will be discussed in Section 4.3.2, one explanation of CMR is that the flanking band permits the listener to detect a change in correlation when the signal is added to the target band. Changing the level of the target band does not alter the correlation between it and the flanking band. Hall, Grose, and Haggard (1990) varied the number of flanking bands from one to eight. The largest CMR occurred with the largest number of flanking bands, but the flanking bands nearest the signal band contributed most heavily to that effect. If the envelopes of two or more of the flanking bands were not correlated with the signal band (deviant flanking bands), the CMR was diminished. The effects of deviant bands were largest when they were located near the signal band. Grose and Hall (1990) showed that uncertainty about the signal location in one of several masking bands had little effect on the measured CMR.

Wright and McFadden (1990) investigated the effects of uncertainty about the correlation between the masking and flanking bands. If the process of detecting the signal is different in the comodulation and reference conditions, then mixing trials with correlated and uncorrelated flanking bands may impede the detection of the signal. Their measurements revealed no change in thresholds when the listeners were uncertain about the masking condition present on any single trial—flanking bands were either correlated or uncorrelated on both presentations of the forced-choice trial. They did, however, discover a potential artifact when different types of stimuli were mixed within the presentations of the forced-choice test. On any presentation, if the flanking bands are correlated, there is then a strong tendency to select as "signal-like" an uncorrelated target band whether it contains a signal or not. This produces a bias in the threshold estimate for those conditions where different types of maskers are presented within the same trial. This bias is also probably important in the comodulation detection difference (see Section 4.4) in which correlated flanking bands elevate the threshold of a correlated target band.

4.3.1.5 Bandwidth of the Noise Band and Amplitude Modulation (AM)

Decreasing the noise bandwidth will, of course, decrease the average rate of envelope modulation. Thus, bandwidth would appear to be a parameter of major importance in CMR research. The only explicit study of this variable was carried out by Schooneveldt and Moore (1987). They studied bandwidths of 25, 50, and 100 Hz, using target and flanking bands at a variety of different frequency spacings. Only the CMRs measured when the center frequencies of the two bands were within one-tenth of an octave showed a notable effect of bandwidth. These results, they argued, were not "true CMRs" because they probably reflected comparisons based on within-band cues rather than across-band cues (see Section 4.3.2). The average CMR (over all frequency spacings) showed a remarkably small change with noise bandwidths. In a more recent paper, Moore and Schooneveldt (1990) used bandwidths of 6.25, 25, and 100 Hz. They calculated what they call a "true" CMR and found values of 0, 4, and 5.6 dB at bandwidths of 100 Hz, 25 Hz, and 6.25 Hz, respectively. Thus, it would appear that CMR does depend on bandwidth, but only at very small (BW < 25 Hz) values. Historically, there has been a persistent trend from larger bandwidths in the early studies to smaller noise bandwidths in more recent experiments. It would be hard to understand this trend if the size of the CMR were really independent of bandwidth.

If a slowly oscillating, well-modulated envelope is the necessary and sufficient condition for the CMR result, then one might expect that amplitude-modulated tones would produce a measurable masking release. Grose and Hall (1989) have shown that such is the case. Moore, Glasberg, and Schooneveldt (1990) have confirmed their results but have also suggested that some of the masking results were produced by what they call "across-channel masking" (ACM). What this phrase means is that some interference (masking) effect occurred for frequency channels that were widely separated in frequency. From an energy standpoint, the bands should be independent. Some of these effects were like those described by Yost and Sheft (1989) and will be discussed in Section 4.4.

4.3.1.6 CMR and Frequency Modulation (FM)

Schooneveldt and Moore (1988) attempted but failed to produce any CMR using frequency—rather than amplitude—modulation of the signal and flanking band. Grose and Hall (1990) also attempted to obtain a CMR using coherent frequency modulation between three bands located at 1600, 2000, and 2400 Hz. The signal, either a pure tone or a band of noise, was added to the middle band. They did find a very small improvement in the audibility of the signal if the coherent FM condition was compared with an incoherent FM condition, but then only for the monaural listening condition. The signal threshold was never smaller in the coherent case than that obtained with the 2000-Hz masking band alone. Therefore, it would not qualify as a CMR by some definitions. Contralateral presentation of the flanking bands was ineffective in all cases. Thus, although only a limited range of frequency modulation parameters have been explored, it appears that CMR effects are much larger with stimuli of time-varying intensity level, such as amplitude-modulated signals.

4.3.2 CMR Theory

There are no detailed quantitative models of CMR; rather, the theories are of a general, qualitative nature. The issues involve the kind of auditory processes that may be responsible for CMR and the types of auditory cues used in those processes. Here is a brief review of the major hypotheses.

The title of the initial paper (Hall Haggard & Fernandes, 1984) suggests that the major cues are a comparison of the temporal patterns from different frequency regions. The nature of the comparison was not specified, but presumably the envelope was extracted both at the frequency of the signal band and at the frequency of the flanking band. Because the two envelopes were correlated in one case and independent waveforms in the other, the signal was more detectable in the former case than the latter. This general hypothesis may be broken into two/three more specific hypotheses about the comparison process.

One idea is that the envelope in the flanking band is used to indicate when, in time, the signal should be particularly detectable in noise. Greater relevance should be given to those times when the envelope is relatively low because the signal-to-noise ratio is then correspondingly high. This idea apparently originated with Buus (1985) and is sometimes called the dip-listening model (or valley-listening model).

A second notion (Buus 1985; Hall 1986) is an adaptation of the binaural model of equalization and cancellation (EC) first suggested by Durlach (1963). As applied to the CMR situation, it assumes that some kind of disparity between the signal and flanking envelopes is the critical decision variable. A simple decision procedure is to subtract the two envelope waveforms. The residue is large when the bands are uncorrelated and practically zero when correlated. Thus, the signal-to-noise ratio is greatly enhanced in the correlated case. This is a completely quantitative model and has only two parameters—an amplitude and a temporal jitter value that prevent perfect cancellation. Despite the quantitative nature of the model, I could find no estimates of these parameters.

A third potential model was first suggested by Richards (1987) who showed that listeners could discriminate monaural envelopes having different degrees of correlation. Discrimination of such changes is more accurate when the initial correlation is perfect ($+$ or -1) than when the correlation is near zero. Thus, the change in envelope correlation produced by adding a signal to two uncorrelated envelopes is less detectable than the change produced by adding a signal to two correlated envelopes. This general idea is similar to the equalization and cancellation model in that linear operations on the two envelopes are the putative decision procedures.

A second class of explanations assumes that the listener monitors the waveform of the signal and flanking band in a single channel, presumably located at some intermediate frequency. This single waveform has a beating pattern whose rate of fluctuation is the difference in frequency of the two bands. The nature of the beating pattern is quite different if the two bands have correlated or uncorrelated envelopes. Figure 2.6 illustrates the summed waveform in the two cases. The signal when added to one of the correlated bands may be more detectable than when added to an uncorrelated band.

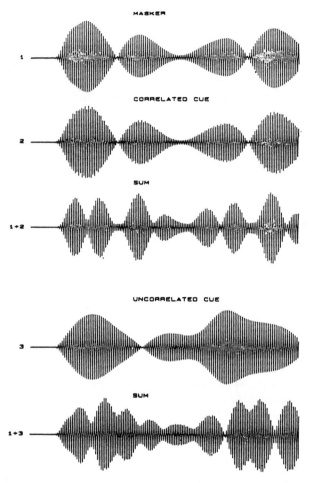

FIGURE 2.6. Example of the summed waveform in a typical CMR experiment. The top three traces show: (1) masker waveform; (2) a correlated cue waveform—the fine structure is slightly different, but the envelope is the same as the masker waveform; and (1 + 2) the sum of these two waveforms. The bottom two traces show: (3) an uncorrelated cue waveform—its envelope is unrelated to the masker waveform; and (1 + 3) the sum of masker and the unrelated cue waveform. The noise bands are 100-Hz wide and centered at 3700 and 3900 Hz. (Taken from McFadden (1986) with permission.)

This general hypothesis is sometimes called simply the "single-band" mechanism. A good review of the envelope properties of a waveform constructed by summing two coherent noise bands is provided by Richards (1987). That such a mechanism may play a role in CMR results was first stated by McFadden (1986) who judiciously phrased it, "it is imprecise to think that one spectral (or basilar membrane) region is receiving only the cue band,

and a second region is receiving only the masker band and the signal, and that the envelopes in the two regions are compared in some way during the process of detection." Such a mechanism has been carefully considered by Moore and Glasberg (1987) in connection with the dip-listening model and by Schooneveldt and Moore (1987, 1989) in their papers on CMR. Clearly, the single-band mechanism may be responsible for at least some of the CMR results, especially for those obtained when the center frequencies of the signal and the flanking band are very near each other. It has never been proposed as the entire basis of CMR. After all, CMR can occur when the flanking and signal bands are in opposite ears. By the principle of Occam's razor, the single-band mechanism should be considered before more elaborate theories are considered.

There is no consensus as to the most probable CMR mechanism. It may well be that, as Hall and Grose (1988) suggested, each explanation has something to offer in different experimental situations.

4.4 Comodulation Detection Difference (CDD)

Shortly after the CMR results became known, Cohen and Schubert (1987b) and McFadden (1987) described an effect that is probably related to CMR and which has become known as comodulation detection difference (CDD). They measured the *detection* of a narrowband noise (target) as a function of the envelope correlation of another narrowband noise (flanker). Detecting the target band was more difficult if the flanker band had the same envelope rather than an unrelated envelope. Note that the CDD effect can be regarded as the opposite of the CMR effect. In CMR, a correlated flanker band makes the signal easier to hear; in CDD, the correlated flanker band makes the signal band more difficult to hear. McFadden (1987) tested the same subjects in the experimental paradigms appropriate for both CMR and CDD. The changes in thresholds were, in fact, in opposite directions for the same listeners tested in the two paradigms.

The difference between the center frequency of the target and flanking bands is an important experimental parameter. Cohen and Schubert (1987a) found the largest effects (about 10 dB) when the flanking band was about an octave above the 1000-Hz target band. They found little effect when the center frequency of the flanking band was below the center frequency of the target band. McFadden (1987), who studied a somewhat higher target band (2500 Hz), found a somewhat smaller CDD that was less frequency dependent but was variable for different listeners.

Wright (1990) explored how these effects changed as the number of signal bands, the number of flanker bands, and the coherence among their envelopes were varied. Using eight flanking bands, the CDD effect was about 7 dB in size; if five signal bands were used, the CDD effect decreased to about 3 dB. In the coherent condition, the envelopes of all bands (signal and flanking) were all the same. In the incoherent condition, the envelopes of all bands were unrelated. It was easier to detect the signal as the number of signal bands increased, but the standard models used to predict such increases were very inaccurate. A dramatic learning effect was evident in three of seven listeners, indicating the need for considerable practice for some listeners in this type of experiment.

McFadden and Wright (1990) have also demonstrated large differences among observers in the way they respond to differences in the relative onsets of the flanking band and the signal band. Using a short (50 ms) signal band, the flanking bands either began simultaneously with the signal band or preceded it by as much as 700 ms. The thresholds for some listeners were hardly affected by this manipulation, while for others the thresholds changed nearly 30 dB (the simultaneous conditions being the most difficult). The change in threshold as a function of this asynchrony was nearly the same and independent of the coherence between the signal and four flanking bands. For a longer duration signal (240 ms), the effects of the relative onset of signal and flanking bands were diminished, but the CDD effects were on average larger. Wright and McFadden (1990) also studied the effects of signal uncertainty in both a CDD and a CMR task. They concluded that uncertainty about the waveform type (correlated envelopes or not) has little effect on detection performance in either of these tasks, although an important potential artifact was discovered (see Section 4.3.1.4).

4.5 Modulation Detection Interference (MDI)

As stated, this review is brief as this topic is treated elsewhere (see Wightman and Kistler, Chapter 5; Yost and Sheft, Chapter 6). The essentials are reviewed so that the results of this interference can be contrasted with the results of the other paradigms. Yost and Sheft (1989) studied the threshold for detecting amplitude modulation of a pure tone in the presence of an interfering tone. The critical stimulus, the target, can be expressed as

$$(1 + m \sin 2\pi f_m t) \cos(2\pi f_t t)$$

where f_m is the modulation frequency, f_t is the target or carrier frequency, and m is the depth of modulation. The discrimination problem is to decide if modulation is present, ($m = 0$ or $m > 0$). Listeners are reasonably sensitive to amplitude modulation at low rates of modulation as Zwicker (1952) showed some years ago. Yost and Sheft (1989, 1990) typically used 5 to 20 Hz as a modulation frequency and found thresholds of about -25 dB ($20 \log m$), a variation in the amplitude of the carrier of approximately 5% can be reliably detected.

The ability to detect amplitude modulation of the target tone is severely impaired if another interfering tone is modulated at the same rate as the target tone. For typical listeners, the threshold value for modulation at the target frequency is elevated by 10 to 15 dB. This interference is caused by *modulation* because an unmodulated flanking tone does not affect the target tone's threshold of amplitude modulation. This is called the modulation detection interference (MDI).

In most experiments, the target and flanker tones are widely separated in frequency; 1000 and 4000 Hz are typical values. If one varies the depth of modulation of the flanking tone, then the interference effect is nearly constant when the depth of the flanker modulation is varied from 100% to 40%, but it goes to zero when no flanker modulation is present (Yost and Sheft 1989). The MDI effect shows a peak of interference when the modulation frequency of the target and flanker are the same. Changing the flanker modulation rate, while holding the target modulation rate constant, produces less interference (Yost, Sheft, and Opie 1989). The degree of tuning is relatively broad; when the two modulation frequencies differ by an octave, the interference effect decreases by only 3 to 6 dB.

A most surprising result is that the amount of interference is practically independent of the phase relation between the target and flanker envelope (Yost and Sheft 1989). This result is particularly awkward, since the authors discuss these results in terms of perceptual groupings and "object" perception. A simple argument is that the common modulation frequency causes the target and flanker frequencies to be heard as a single auditory object, thus making it more difficult to determine a single feature of the target sound. One may question the principles that govern perceptual grouping. Are sounds grouped together because they have a common modulation frequency, even if sound A waxes while sound B wanes and vice versa?

Yost, Sheft, and Opie (1989) also showed that the amount of interference depends on the rate of amplitude modulation. When both tones are modulated at the same rate, the interference effect is greatest at the lowest rate of modulation, about 2 Hz, and the effect appears to decrease monotonically, with essentially no interference at about 100-Hz modulation rate. In the same study, they also show that detection of a change in modulation rate, Δf_m, at the target frequency is impeded by the presence of a modulating flanker tone. In a recent paper, they compared MDI, CMR, and CDD in a single experimental situation (Yost and Sheft 1990). They found modulation detection interference but did not find comodulation masking release or comodulation detection difference.

Acknowledgements. This research was supported by a grants from the Air Force Office of Scientific Research and the National Institutes of Health. I wish to thank Dr. Haunping Dai for his extensive comments. Ms. Mary Fullerton organized and cross-checked the references. Ms. Marian Green edited several drafts of the chapter. The author also wishes to express his sincere thanks to Dr. Joseph Hall and Dr. Dennis McFadden who corrected many errors of interpretation and expression in previous drafts of this chapter.

References

Berg BG (1990) Analysis of weights in multiple observation tasks. J Acoust Soc Am 86:1743–1746

Berg BG (1991) paper in preparation.

Berg BG, Green DM (1990) Spectral weights in profile listening. J Acoust Soc Am 88:758–766.

Berliner JE, Durlach NI (1973) Intensity perception. IV. Resolution in roving-level discrimination. J Acoust Soc Am 53:1270–1287.

Berliner JE, Durlach NI, Braida LD (1977) Intensity perception. VII. Further data on roving-level discrimination and the resolution and bias edge effects. J Acoust Soc Am 61:1577–1585.

Bernstein LB, Green DM (1988) Detection of changes in spectral shape: Uniform vs. non-uniform background spectra. Hear Res 32:157–165.

Buus S (1985) Release from masking caused by envelope fluctuations. J Acoust Soc Am 78:1958–1965.

Buus S, Florentine M (1989) Psychometric functions for level discrimination. Unpublished manuscript.

Cohen MF, Schubert ED (1987a) Influence of place synchrony on detection of a sinusoid. J Acoust Soc Am 81:452–458.

Cohen MF, Schubert ED (1987b) The effect of cross-spectrum correlation on the detectability of a noise band. J Acoust Soc Am 81:721–723.

Durlach NI (1963) Equalization and cancellation theory of binaural masking level differences. J Acoust Soc Am 35:1206–1218.

Durlach NI, Braida LD, Ito Y (1986) Toward a model for discrimination of broadband signals. J Acoust Soc Am 80:63–72.

Feth LL (1974) Frequency discrimination of complex periodic tones. Percept Psychophys 15:375–378.

Fletcher H (1940) Auditory patterns. Rev Mod Phys 12:47–65.

Fletcher H (1953) Speech and Hearing in Communication, Second Edition. New York: D. Van Nostrand, (First Edition 1929).

Florentine M (1983) Intensity discrimination as a function of level and frequency and its relation to high-frequency hearing. J Acoust Soc Am 74:1375–1379.

Florentine M (1986) Level discrimination of tones as a function of duration. J Acoust Soc Am 79:792–798.

Garner WR, Miller GA (1944) Differential sensitivity to intensity as a function of the duration of the comparison tone. J Exp Psychol 34:450–463.

Grantham DW, Yost WA (1982) Measures of intensity discrimination. J Acoust Soc Am 72:406–410.

Green DM (1958) Detection of multiple component signals in noise. J Acoust Soc Am 30:904–911.

Green DM (1960) Psychoacoustics and detection theory. J Acoust Soc Am 31:1189–1202.

Green DM (1983) Profile Analysis: A different view of auditory intensity discrimination. Am Psychol 38:133–142.

Green DM (1988) Profile Analysis; Auditory Intensity Discrimination. New York, Oxford: Oxford University Press.

Green DM, Birdsall TG (1978) Detection and recognition. Psychol Rev 85:192–206.

Green DM, Dai H (1991) Probability of being correct with 1 of M orthogonal signals. Notes and Comment, Percept Psychophys 49:100–101.

Green DM, Mason CR (1985) Auditory profile analysis: Frequency, phase, and Weber's Law. J Acoust Soc Am 77:1155–1161.

Green DM, Sewall S (1962) Effects of background noise on auditory detection of noise bursts. J Acoust Soc Am 34:1217–1223.

Green DM, Swets JA (1966) Signal Detection Theory and Psychophysics. New York: John Wiley and Sons (reprinted Huntington, NY: Robert E. Krieger, 1974; Los Altos, CA: Peninsula Publishing, 1988).

Green DM, Birdsall TG, Tanner WP Jr (1957) Signal detection as a function of signal intensity and duration. J Acoust Soc Am 29:523–531.

Green DM, Licklider JCR, McKey MJ (1959) Detection of pulsed sinusoid in noise as a function of frequency. J Acoust Soc Am 31:1446–1452.

Green DM, Nachmias J, Kearney JK, Jeffress LA (1979) Intensity discrimination with gated and continuous sinusoids. J Acoust Soc Am 66:1051–1056.

Green DM, Kidd G, Jr, Picardi MC (1983) Successive versus simultaneous comparison in auditory intensity discrimination. J Acoust Soc Am 73:639–643.

Green DM, Onsan ZA, Forrest TG (1987) Frequency effects in profile analysis and detecting complex spectra changes. J Acoust Soc Am 81:692–699.

Grose JH, Hall JW, III (1989) Comodulation masking release using SAM tonal complex maskers: Effects of modulation depth and signal position. J Acoust Soc Am 85:1276–1284.

Grose JH, Hall JW III (1990) The effect of signal-frequency uncertainty on comodulation masking release. J Acoust Soc Am 87:1272–1277.

Hall JW III (1986) The effect of across-frequency differences in masking level on spectro-temporal pattern analysis. J Acoust Soc Am 79:781–787.

Hall JW III, Grose JH (1988) Comodulation masking release: Evidence for multiple cues. J Acoust Soc Am 84:1669–1675.

Hall JW III, Haggard MP, Fernandes MA (1984) Detection in noise by spectro-temporal pattern analysis. J Acoust Soc Am 76:50–56.

Hall JW III, Grose JH, Haggard MP (1988) Comodulation masking release for multicomponent signals. J Acoust Soc Am 83:677–686.

Hall JW III, Grose JH, Haggard MP (1990) Effects of flanking band proximity, number, and modulation pattern on comodulation masking release. J Acoust Soc Am 87:269–283.

Hanna TE, von Gierke SM, Green DM (1986) Detection and intensity discrimination of a sinusoid. J Acoust Soc Am 80:1335–1340.

Hawkins JE, Stevens SS (1950) The masking of pure tones and of speech by white noise. J Acoust Soc Am 22:6–13.

Henning GB (1970) Comparison of the effects of signal duration on frequency and amplitude discrimination. In: Plomp R, Smoorenburg GF (eds) Frequency Analysis and Periodicity Detection in Hearing Leiden, The Netherlands: AW Sijthoff, pp. 350–361.

Jesteadt W, Wier CC, Green DM (1977) Intensity discrimination as a function of frequency and sensation level. J Acoust Soc Am 61:169–177.

Laming D (1986) Sensory Analysis. London: Academic Press, Harcourt Brace Jovanovich Publishers.

Marill TM (1956) Detection Theory and Psychophysics. Technical Report No. 319, Research Laboratories of Electronics, Massachusetts Institute of Technology, Cambridge.

McFadden DM (1986) Comodulation masking release: Effects of varying the level, duration, and time delay of the cue band. J Acoust Soc Am 80:1658–1667.

McFadden DM (1987) Comodulation detection differences using noise-band signals. J Acoust Soc Am 81:1519–1527.

McFadden DM, Wright BA (1990) Temporal decline of masking and comodulation detection differences. J Acoust Soc Am 88:711–724.

McGill WJ, Goldberg JP (1968) A study of the near-miss involving Weber's Law and pure-tone intensity discrimination. Percept Psychophys 4:105–109.

Middleton D (1954) Statistical theory of signal detection. Professional Group on Information Theory, PGIT-7, pp 105–113.

Moore BCJ, Glasberg BR (1987) Factors affecting thresholds for sinusoidal signals in narrow-band maskers with fluctuating envelopes. J Acoust Soc Am 82:69–79.

Moore BCJ, Glasberg BR, Schooneveldt GP (1990) Across-channel masking and comodulation masking release. J Acoust Soc Am 87:1683–1694.

Moore BCJ, Hall JW III, Grose JH, Schooneveldt GP (1990) Some factors affecting the magnitude of comodulation masking release. J Acoust Soc Am 88:1694–1702.

Penner MJ, Leshowitz B, Cudahy E, Ricard G (1974) Intensity discrimination for pulsed sinusoids of various frequencies. Percept Psychophys 15:568–570.

Peterson WW, Birdsall TG, Fox WC (1954) The theory of signal detectability. Professional Group on Information Theory. PGIT-4, pp. 171–212.

Pfafflin SM, Mathews MV (1962) Energy detection model for monaural auditory detection. J Acoust Soc Am 34:1842–1852.

Pillsbury WB (1910) Method for the determination of the intensity of sound in psychological monographs. Report of the Committee of the American Psychological Association on the Standardizing of Procedures in Experimental Tests, XIII, pp. 5–20.

Plomp R, Bouman MA (1959) Relation between hearing threshold and duration for tone pulses. J Acoust Soc Am 31:749–758.

Pollack I (1955) "Long-time" differential intensity sensitivity. J Acoust Soc Am 27:380–381.

Pynn CT, Braida LD, Durlach NI (1972) Intensity perception. III. Resolution in small-range identification. J Acoust Soc Am 51:559–566.

Rabinowitz WM, Lim JS, Braida LD, Durlach NI (1976) Intensity perception. VI. Summary of recent data on deviations from Weber's Law for 1000-Hz tone pulses. J Acoust Soc Am 59:1506–1509.

Rayleigh Baron (Strutt, JW) (1945) Theory of Sound, 1, New York: Dover, (Originally published 1877).

Richards VM (1987) Monaural envelope correlation perception. J Acoust Soc Am 82:1621–1630.

Riesz RR (1928) Differential sensitivity of the ear for pure tones. Phys Rev 31:867–875.

Schacknow PN, Raab DH (1973) Intensity discrimination of tone bursts and the form of the Weber function. Percept Psychophys 14:449–450.

Scharf B (1970) Critical bands. Foundations of Modern Auditory Theory, New York, London: Academic Press 1:159–202.

Schooneveldt GP, Moore BCJ (1987) Comodulation masking release (CMR): Effects of signal frequency, flanking band frequency, masker bandwidth, flanking band level, monotic versus dichotic presentation of the flanking band. J Acoust Soc Am 82:1944–1956.

Schooneveldt GP, Moore BCJ (1988) Failure to obtain comodulation masking release with frequency-modulated maskers. J Acoust Soc Am 83:2290–2293.

Schooneveldt GP, Moore BCJ (1989) Comodulation masking release (CMR) for various monaural and binaural combinations of the signal, on-frequency, and flanking bands. J Acoust Soc Am 85:262–272.

Sorkin RD (1966) Temporal interference effects in auditory amplitude discrimination. Percept Psychophys 1:55–58.

Tanner WP, Jr (1961) Physiological implications of psychophysical data. Ann NY Acad Sci 89:752–765.

Viemeister NF (1972) Intensity discrimination of pulsed sinusoids: The effects of filtered noise. J Acoust Soc Am 51:1265–1269.

Watson CS, Gengel RW (1969) Signal duration and signal frequency in relation to auditory sensitivity. J Acoust Soc Am 46:989–996.

Watson CS, Franks JR, Hood DC (1972) Detection of tones in the absence of external masking noise. I. Effects of signal intensity and signal frequency. J Acoust Soc Am 52:633–643.

Wright BA (1990) Comodulation detection differences with multiple signal bands. J Acoust Soc Am 87:292–303.

Wright BA, McFadden D (1990) Uncertainty about the correlation among temporal envelopes in two comodulation tasks. J Acoust Soc Am 88:1339–1350.

Yost WA, Sheft S (1989) Across-critical-band processing of amplitude-modulated tones. J Acoust Soc Am 85:848–857.

Yost WA, Sheft S (1990) A comparison among three measures of cross-spectral processing of amplitude modulation with tonal signals. Letters to the Editor, J Acoust Soc Am 87:897–900.

Yost WA, Sheft S, Opie J (1989) Modulation interference in detection and discrimination of amplitude modulation. J Acoust Soc Am 86:2138–2147.

Zwicker E (1952) Die Grenzen der Horbarkeit der Amplitudenmodulation und der Frequenzmodulation eines Tones. Acustica 2, Beifeft 3, AB 125.

Zwicker E (1953) Uber die Horbarkeit nichtsinusformiger Tonhohenschwankungen, Fund und Ton 7:342.

Zwicker E, Henning GB (1985) The four factors leading to binaural masking-level differences. Hear Res 19:29–47.

Zwicker E, Wright HN (1963) Temporal summation for tones in narrow-band noise. J Acoust Soc Am 35:691–699.

Zwicker E, Flottorp G, Stevens SS (1957) Critical band width in loudness summation. J Acoust Soc Am 29:548–557.

3
Frequency Analysis and Pitch Perception

BRIAN C.J. MOORE

1. Introduction

This chapter is concerned with two main areas: frequency analysis and pitch perception. Frequency analysis refers to the action of the ear in resolving (to a limited extent) the sinusoidal components in a complex sound; this ability is also known as frequency selectivity and frequency resolution. It plays a role in many aspects of auditory perception but is most often demonstrated and measured by studying masking. Studies of pitch perception are mainly concerned with the relationships between the physical properties of sounds and the perceived pitches of those sounds and with the underlying mechanisms that explain these relationships. One important aspect of pitch perception is frequency discrimination, which refers to the ability to detect changes in frequency over time and which is (at least partly) a separate ability from frequency selectivity.

2. The Power Spectrum Model and the Concept of the Critical Band

Fletcher (1940) measured the threshold for detecting a sinusoidal signal as a function of the bandwidth of a bandpass noise masker. The noise was always centered at the signal frequency, and the noise power density was held constant. Thus, the total noise power increased as the bandwidth increased. This experiment has been repeated several times since then (Hamilton 1957; Greenwood 1961; Spiegel 1981; Schooneveldt and Moore 1989). An example of the results, taken from Schooneveldt and Moore, is given in Figure 3.1. The threshold of the signal increases at first as the noise bandwidth increases but then flattens off, so that further increases in noise bandwidth do not change the signal threshold significantly.

To account for these results, Fletcher (1940) suggested that the peripheral auditory system behaves as if it contained a bank of bandpass filters, with

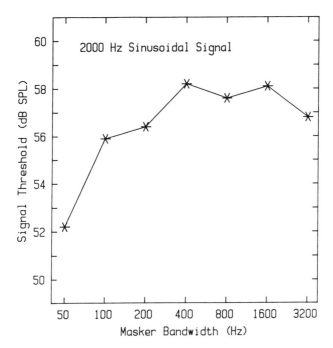

FIGURE 3.1. The threshold of a 2000-Hz sinusoidal signal plotted as a function of the bandwidth of a noise masker centered at 2000 Hz. (From Schooneveldt and Moore 1989.)

overlapping passbands. These filters are now called the "auditory filters." Fletcher suggested that the signal was detected by attending to the output of the auditory filter centered on the signal frequency. Increases in noise bandwidth result in more noise passing through that filter, provided the noise bandwidth is less than the filter bandwidth. However, once the noise bandwidth exceeds the filter bandwidth, further increases in noise bandwidth will not increase the noise passing through the filter. Fletcher called the bandwidth at which the signal threshold ceased to increase the "critical bandwidth" (*CB*).

Fletcher's experiment led to a model of masking known as the power-spectrum model that is based on the following assumptions:

(1) The peripheral auditory system contains an array of linear overlapping bandpass filters.
(2) When trying to detect a signal in a noise background, the listener is assumed to make use of just one filter with a center frequency close to that of the signal. This filter will pass the signal but remove a great deal of the noise.
(3) Only the components in the noise which pass through the filter will have any effect in masking the signal.

(4) The threshold for detecting the signal is determined by the amount of noise passing through the auditory filter; specifically, the threshold is assumed to correspond to a certain signal-to-noise ratio, K, at the output of the filter. The stimuli are represented by their long-term power spectra, i.e., the relative phases of the components and the short-term fluctuations in the masker are ignored.

It is now known that none of these assumptions is strictly correct. In particular, the filters are not strictly linear (e.g., Moore and Glasberg 1987); listeners can combine information from more than one filter to enhance signal detection (Spiegel 1981; Buus et al. 1986); noise falling outside the passband of the auditory filter centered at the signal frequency can affect the detection of that signal (Hall, Haggard, and Fernandes 1984), and fluctuations in the masker can play a strong role (Patterson and Henning 1977). Nevertheless, the concept of the auditory filter is widely accepted and very useful, and departures from linearity can be small if the range of noise levels at the input is small. Further, although the assumptions of the model do sometimes fail, it works well in many situations, provided the stimuli are carefully chosen.

In analyzing the results of his experiment, Fletcher (1940) made a simplifying assumption. He assumed that the shape of the auditory filter could be approximated as a simple rectangle, with a flat top and vertical edges. For such a filter, all components within the passband of the filter are passed equally and all components outside the passband are removed. The width of this passband is equal to the *CB* described above. The term "critical band" is often used to refer to this hypothetical rectangular filter. It should be emphasised that the auditory filter is not rectangular and that the data rarely show a distinct breakpoint corresponding to the *CB*. It is surprising how, even today, many researchers talk about the critical band as if the underlying filter were rectangular.

Fletcher (1940) pointed out that the value of the *CB* could be estimated indirectly by measuring the power of a sinusoidal signal (P_s) required for the signal to be detected in broadband white noise, given the assumptions of the power-spectrum model.

For a white noise with power density, N_o, the total noise power falling within the *CB* is $N_o \times CB$. According to assumption (4) above,

$$P_s/(CB \times N_o) = K$$

and

$$CB = P_s/(K \times N_o).$$

By measuring P_s and N_o and estimating K, the value of the *CB* can be evaluated.

Fletcher (1940) estimated K to equal 1, indicating that the value of the *CB* should be equal to P_s/N_o. The ratio P_s/N_o is now usually known as the critical ratio. Unfortunately, Fletcher's estimate of K has turned out not to be accurate. More recent experiments show that K is typically about 0.4 (Scharf 1970). Thus, at most frequencies, the critical ratio is about 0.4 times the value of the CB estimated by more direct methods such as the band-widening experiment. Also, K varies with center frequency, so the critical ratio does not give a correct indication of how the CB varies with center frequency (Patterson and Moore 1986; Moore, Peters, and Glasberg 1990).

3. Estimating the Shape of the Auditory Filter

Most methods for estimating the shape of the auditory filter at a given center frequency are based on the assumptions of the power-spectrum model of masking. If the masker is represented by its long-term power spectrum, $N(f)$, and the weighting function or shape of the auditory fitter is $W(f)$, then the power-spectrum model is expressed by:

$$P_s = K \int_0^\infty N(f)W(f)\,df, \qquad (1)$$

where P_s is the power of the signal at threshold. By manipulating the masker spectrum, $N(f)$, and measuring the corresponding changes in P_s, it is possible to derive the filter shape, $W(f)$.

The masker chosen to measure the auditory filter shape should be such that the assumptions of the power-spectrum model are not strongly violated (see also Green, Chapter 2). A number of factors affect this choice. If the masker is composed of one or more sinusoids, beats between the signal and masker may provide a cue to the presence of the signal. For small frequency differences between the signal and masker, the beats are perceived as periodic changes in loudness. For greater differences, there is a sensation of roughness or dissonance (Plomp and Steeneken 1968). For very large differences, the beats are not perceptible. This makes sinusoids unsuitable as maskers for estimating the auditory filter shape, since the salience of beats as a cue changes as the masker frequency is altered; this would violate the assumption of the power-spectrum model that threshold corresponds to a constant signal-to-masker ratio at the output of the auditory filter.

In general, noise maskers are more suitable than sinusoids for estimating auditory filter shape, since noises have inherent amplitude fluctuations which make beats much less effective as a cue. However, for narrowband noises which have relatively slow fluctuations, temporal interactions between the signal and masker may still be audible. In addition, the slow fluctuations may strongly influence the detectability of the signal in a way which depends on the difference between the center frequency of the masker and the frequency of the signal (Buus 1985). For these reasons, the assumptions of the power-spectrum model are best satisfied using reasonably broadband noise maskers.

A second important consideration in choosing a noise masker for measuring auditory filter shapes is that the filter giving the highest signal-to-masker ratio is not necessarily centered at the signal frequency. When a listener detects the signal through a filter that is not centered at the signal frequency, this is called "off-frequency listening." Furthermore, if the masker spectrum is concentrated primarily above or below the signal frequency, there may be a range of filter center frequencies over which the signal-to-masker ratio is sufficiently high to give useful information. Under these conditions, the observer may combine information over several auditory filters rather than listening through a single filter, as assumed by the power-spectrum model (Patterson and Moore 1986; for a similar concept applied to intensity discrimination, see Florentine and Buus 1981).

3.1 Psychophysical Tuning Curves

Psychophysical tuning curves (PTCs) are relatively simple to determine. They do, however, suffer from several of the problems described in Section 3. PTCs involve a procedure which is analogous in many ways to the determination of a neural tuning curve (Chistovich 1957; Small 1959). To determine a PTC, the signal is fixed in level, usually a very low level, say, 10 dB SL. The masker can be either a sinusoid or a narrow band of noise, but a noise is generally preferred, for the reasons given in Section 3.

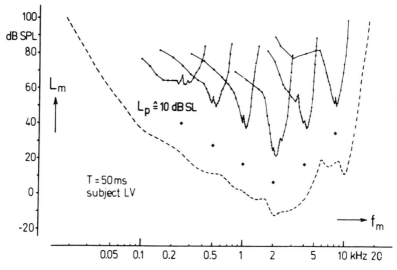

FIGURE 3.2. Psychophysical tuning curves (PTCs) determined in simultaneous masking, using sinusoidal signals at 10 dB sensation level (SL). For each curve, the solid diamond below it indicates the frequency and level of the signal. The masker was a sinusoid which had a fixed starting phase relationship to the brief, 50 ms signal. The masker level, L_m, required for threshold is plotted as a function of masker frequency, f_m, on a logarithmic scale. The dashed line shows the absolute threshold for the signal. (From Vogten, 1974, by permission of the author.)

For each of several masker center frequencies, the level of the masker needed just to mask the signal is determined. Because the signal is at a low level, it is assumed that it will produce activity primarily in one auditory filter. It is assumed further that, at threshold, the masker produces a constant output from that filter in order to mask the fixed signal. Thus, the PTC will indicate the masker level required to produce a fixed output from the auditory filter as a function of frequency. Normally, a filter characteristic is determined by plotting the output from the filter for an input varying in frequency and fixed in level. However, if the filter is linear, the two methods will give the same result. Thus, if linearity is assumed, the shape of the auditory filter can be obtained simply by inverting the PTC. Examples of some PTCs are given in Figure 3.2.

It has been assumed so far that only one auditory filter is involved in the determination of a PTC. However, there is now good evidence that "off-frequency listening" can influence PTCs. The result of off-frequency listening

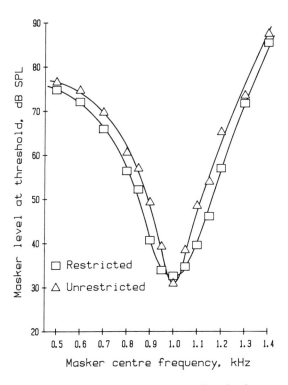

FIGURE 3.3. Comparison of PTCs where off-frequency listening is not restricted (triangles) and where it is restrict (squares) using a low-level notched noise centered at the signal frequency. (Data from Moore, Glasberg, and Roberts 1984.)

is that the PTC has a sharper tip than would be obtained if only one auditory filter were involved (Johnson-Davies and Patterson 1979; O'Loughlin and Moore 1981).

One way to limit off-frequency listening is to add to the masker a fixed, low-level noise with a spectral notch centered at the signal frequency (O'Loughlin and Moore 1981; Moore, Glasberg, and Roberts 1984; Patterson and Moore 1986). Such a masker should make it disadvantageous to use an auditory filter whose center frequency is shifted much from the signal frequency. The effect of using such a noise, in addition to the variable narrowband masker, is illustrated in Figure 3.3. The main effect is to broaden the tip of the PTC; the slopes of the skirts are relatively unaffected.

A final difficulty in using PTCs as a measure of frequency selectivity is connected with the nonlinearity of the auditory filter. Evidence will be presented in Section 4.2 indicating that the auditory filter is not strictly linear but changes its shape with level. The shape seems to depend more on the level at the input to the filter than on the level at the output. However, in determining a PTC, the input is varied while the output is held (roughly) constant. Thus, effectively, the underlying filter shape changes as the masker frequency is altered. This can give a misleading impression of the shape of the auditory filter; in particular, it leads to an underestimation of the slope of the lower skirt of the filter and an overestimation of the slope of the upper skirt (Verschuure 1981; Moore and O'Loughlin 1986).

3.2 The Notched-Noise Method

To satisfy the assumptions of the power-spectrum model, it is necessary to use a masker that limits both the amount by which the center frequency of the filter can be shifted (off-frequency listening) and the range of filter center frequencies over which the signal-to-masker ratio is sufficiently high to be useful. This can be achieved using a noise masker with a spectral notch around the signal frequency. For such a masker, the highest signal-to-masker ratio will occur for a filter which is centered reasonably close to the signal frequency, and performance will not be improved by combining information over filters covering a range of center frequencies (Patterson 1976; Patterson and Moore 1986). The filter shape can then be estimated by measuring signal threshold as a function of the width of the notch.

For moderate noise levels, the auditory filter is almost symmetrical on a linear frequency scale (Patterson 1974, 1976; Patterson and Nimmo-Smith 1980). Hence, the auditory filter shape can be estimated using a notched-noise masker with the notch placed symmetrically about the signal frequency. The method is illustrated in Figure 3.4. For a masker with a notch width of $2\Delta f$ and a center frequency, f_o, Equation 1 becomes:

$$P_s = KN_o \int_0^{f_o - \Delta f} W(f)\,df + KN_o \int_{f_o + \Delta f}^{\infty} W(f)\,df, \qquad (2)$$

3. Frequency Analysis and Pitch Perception   63

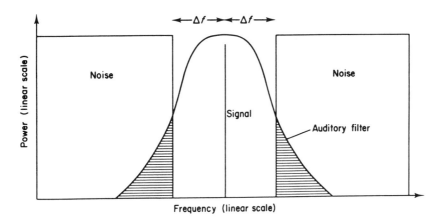

FIGURE 3.4. Schematic illustration of the technique used by Patterson (1976) to determine the shape of the auditory filter. The threshold of the sinusoidal signal is measured as a function of the width of a spectral notch in the noise masker, specified as the deviation, Δf, of the notch edge from the signal frequency. The amount of noise passing through the auditory filter centered at the signal frequency is proportional to the shaded areas.

where N_o is the power spectral density of the noise in its passbands. The two integrals on the right-hand side of Equation 2 represent the respective areas in Figure 3.4 where the lower and upper noise bands overlap the filter. Since both the filter and the masker are symmetrical about the signal frequency, these two areas are equal. Thus, the function relating P_s to the width of the notch provides a measure of the integral of the auditory filter. Hence, the value of $W(f)$ at a given deviation, Δf, from the center frequency is given by the slope of the threshold function at a notch width of $2\Delta f$.

When the auditory filter is asymmetrical, as it is at high masker levels (see Section 4.2), then the filter shape can still be measured using a notched-noise masker, if some reasonable assumptions are made and the range of measurements is extended to include conditions where the notch is placed asymmetrically about the signal frequency. It is first necessary to assume that the auditory filter shape can be approximated by a simple mathematical expression with a small number of free parameters. Patterson et al. (1982) suggested a family of such expressions, all having the form of an exponential with a rounded top, called "roex" for brevity. The simplest of these expressions was called the roex(p) filter shape. It is convenient to measure frequency in terms of the absolute value of the deviation from the center frequency of the filter, f_o, and to normalize this frequency variable by dividing by the center frequency of the filter. The new frequency variable, g, is:

$$g = |f - f_o|/f_o. \qquad (3)$$

The roex(p) filter shape is then given by:

$$W(g) = (1 + pg)\exp(-pg), \qquad (4)$$

where p is a parameter which determines both the bandwidth and the slope of the skirts of the auditory filter. The higher the value of p, the more sharply tuned is the filter. The equivalent rectangular bandwidth (ERB) is equal to $4f_o/p$. When the filter is assumed to be asymmetrical, then p is allowed to have different values on the two sides of the filter: p_l for the lower branch and p_u for the upper branch. The ERB in this case is $2f_o/p_l + 2f_o/p_u$.

Having assumed this general form for the auditory filter shape, the values of p_l and p_u for a particular experiment can be determined by rewriting Equation 2 in terms of the variable g and substituting the above expression for W (Equation 4); the value of p_l is used for the first integral and the value of p_u for the second. The equation can then be solved analytically (for full details see Patterson et al. 1982; Glasberg et al. 1984). Starting values of p_l and p_u are assumed, and the equation is used to predict the threshold for each condition (for notches placed both symmetrically and asymmetrically about the signal frequency). The center frequency of the filter is allowed to shift for each condition so as to find the center frequency giving the highest signal-to-masker ratio; this center frequency is assumed in making the prediction for that condition. Standard least-squares minimization procedures are then used to find the values of p_l and p_u which minimize the mean square deviation between the obtained and predicted values. The minimization is done with the thresholds expressed in decibels. Full details are given in Patterson and Nimmo-Smith (1980), Glasberg et al. (1984), Patterson and Moore (1986) and Glasberg and Moore (1990).

The roex(p) filter shape is usually quite successful in predicting the data from notched-noise experiments, except when the masked thresholds approach absolute threshold. In such cases, there is a decrease in the slope of the function relating threshold to notch width, a decrease which is not predicted by the roex(p) filter shape. This can be accommodated by limiting the dynamic range of the filter using a second parameter, r. This gives the roex(p,r) filter shape of Patterson et al. (1982):

$$W(g) = (1 - r)(1 + pg)\exp(-pg) + r. \qquad (5)$$

As before, p can have different values for the upper and lower branches of the filter. However, the data can be well predicted using the same value of r for the two sides of the filter (Glasberg et al. 1984; Tyler et al. 1984; Glasberg and Moore 1986). The method of deriving filter shapes using this expression is exactly analogous to that described above.

In summary, although the method of deriving auditory filter shapes described above is somewhat complex, it has the advantage of avoiding serious violations of the assumptions of the power-spectrum model. In particular, the detection cues used by the subject do not appear to alter markedly as the spectrum of the masker is manipulated, and the degree of off-frequency listening is controlled and taken into account in the derivation.

3.3 The Rippled-Noise Method

Several researchers have estimated auditory filter shapes using rippled noise, sometimes also called comb-filtered noise, as a masker. This is produced by adding a white noise to a copy of itself which has been delayed by T seconds. The resulting spectrum has peaks spaced at $1/T$ Hz, with minima in between. When the delayed version of the noise is added to the original in phase, the first peak in the spectrum of the noise occurs at 0 Hz; this noise is referred to as cosine+. When the polarity of the delayed noise is reversed, the first peak is at $0.5/T$ Hz; this is referred to as cosine−. The sinusoidal signal is usually fixed in frequency, and the values of T are chosen so that the signal falls at either a maximum or a minimum in the masker spectrum; the signal threshold is measured for both cosine+ and cosine− noise for various ripple densities (different values of T).

The auditory filter shape can be derived from the data either by approximating the auditory filter as a Fourier series (Houtgast 1977; Pick 1980) or by a method similar to that described for the notched-noise method (Glasberg, Moore, and Nimmo-Smith 1984; for a review see Patterson and Moore 1986). The filter shapes obtained in this way are generally similar to those obtained using the notched-noise method, although they tend to have a slightly broader and flatter top (Glasberg, Moore, and Nimmo-Smith 1984). The method does not allow the auditory filter shape to be measured over a wide dynamic range.

3.4 Allowing for the Transfer Function of the Outer and Middle Ear

The transfer function of the outer and middle ear varies markedly with frequency, particularly at very low and high center frequencies. Clearly, this can have a significant influence on measures of frequency selectivity. For example, if one of the bands of noise in a notched-noise experiment is very low or high in center frequency, it will be strongly attenuated by the middle ear and so will not do much masking. The auditory filter is usually conceived of as resulting from processes occurring after the middle ear. Hence, the effect of the outer and middle ears could be conceived as a frequency-dependent attenuation applied to all stimuli before auditory filtering takes place. If this is the case, then the frequency-dependent attenuation should be taken into account in the fitting procedure for deriving filter shapes. Essentially, the spectra of the stimuli at the input to the cochlea have to be calculated by assuming a certain form for the frequency-dependent transfer. The fitting procedure then has to work on the basis of these "corrected" spectra. In practice, this means that the integral in equation 2 cannot be solved analytically, but has to be evaluated numerically. Glasberg and Moore (1990) have considered three possible types of "correction" and have suggested appropriate conditions under which each one should be used. They list a computer

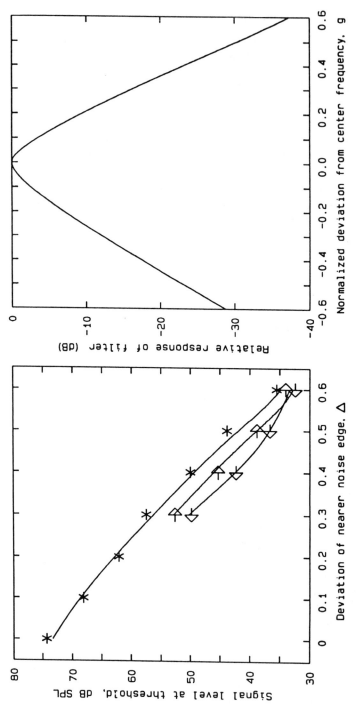

FIGURE 3.5. The left panel shows thresholds for a 200-Hz signal as a function of the width of a notch in a noise masker. The value on the abscissa is the deviation of the nearer edge of the notch from the signal frequency, divided by the signal frequency, represented by the symbol Δ. Asterisks (*) indicate conditions where the notch was placed symmetrically about the signal frequency. Right-pointing arrows indicate conditions where the upper edge of the notch was 0.2 units farther away from the signal frequency than the lower edge. Left-pointing arrows indicate conditions where the lower edge of the notch was 0.2 units farther away than the upper edge. The fact that the left-pointing arrows are markedly below the right-pointing arrows indicates that the filter is asymmetrical. The right panel shows the auditory filter shape derived from the data. (From Moore, Peters, and Glasberg 1990.)

program for deriving auditory filter shapes from notched-noise data that includes the option of using "corrections" to allow for the transfer function of the outer and middle ear.

3.5 An Example of Measurement of the Auditory Filter Shape

In this section, an example of data obtained using the notched-noise method and the filter shape obtained is given. The data are for a normally hearing subject and a signal frequency of 200 Hz. In the left panel of Figure 3.5, signal thresholds are plotted as a function of the width of the spectral notch in the noise masker. The lines in the left panel are the fitted values derived from the roex(p,r) model, using the "ELC" (Equal Loudness Contour) correction described by Glasberg and Moore (1990). The model fits the data well. The derived filter shape is shown in the right panel of Figure 3.5. The filter is somewhat asymmetric, with a shallower lower branch. The ERB is 48 Hz, which is typical at this center frequency.

4. Summary of the Characteristics of the Auditory Filter

4.1 Variation with Center Frequency

Moore and Glasberg (1983) presented a summary of experiments measuring auditory filter shapes using symmetrical notched-noise maskers. All of the data were obtained at moderate noise levels and were analyzed using the roex(p,r) filter shape. More recently, Glasberg and Moore (1990) have updated that summary, including results that extend the frequency range of the measurements and data from experiments using asymmetrical notches. The ERBs of the filters derived from the data available in 1983 are shown as asterisks in Figure 3.6. The dashed line shows the equation fitted to the data in 1983. Other symbols show ERBs estimated in more recent experiments, as indicated in the figure.

The solid line in Figure 3.6 provides a good fit to the ERB values over the whole frequency range tested. It is described by the following equation:

$$\text{ERB} = 24.7(4.37F + 1), \qquad (6)$$

where F is frequency in kHz. This equation is a modification of one originally suggested by Greenwood (1961) to describe the variation of the CB with center frequency. He based it on the assumption that each CB corresponds to a constant distance along the basilar membrane. Although the constants in Equation 6 differ from those given by Greenwood, the form of the equation is the same as his. This is consistent with the idea that each ERB corresponds to a constant distance along the basilar membrane (Moore 1986).

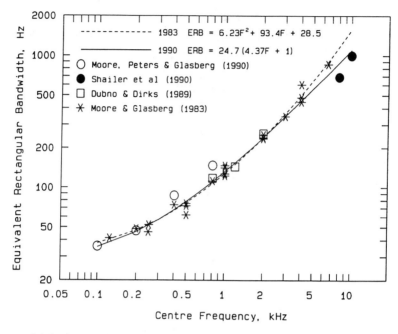

FIGURE 3.6. Estimates of the equivalent rectangular bandwidth (ERB) of the auditory filter from a variety of experiments, plotted as a function of center frequency (f). The dashed line represents the equation suggested by Moore and Glasberg (1983). The solid line represents the equation suggested by Glasberg and Moore (1990). (Adapted from Glasberg and Moore 1990.)

It should be noted that the function specified by Equation 6 differs somewhat from the "traditional" critical band function (Zwicker 1961) which flattens off below 500 Hz at a value of about 100 Hz. The traditional function was obtained by combining data from a variety of experiments involving masking, loudness, and phase sensitivity (see Section 6 for further details of these other measures). However, the data were sparse at low frequencies, and the form of the function at low frequencies was strongly influenced by measures of the critical ratio. As described in Section 2, the critical ratio does not provide a good estimate of the CB, particularly at low frequencies (Moore, Peters, and Glasberg 1990). It seems clear that the CB does continue to decrease below 500 Hz.

Auditory filter shapes for young, normally hearing subjects vary relatively little among the subjects; the standard deviation of the ERB is typically about 10% of its mean value (Moore 1987; Moore, Peters, and Glasberg 1990). However, the variability tends to increase at very low frequencies (Moore, Peters, and Glasberg 1990) and at very high frequencies (Patterson et al. 1982; Shailer et al. 1990).

4.2 The Variation of the Auditory Filter Shape with Level

If the auditory filter was linear, then its shape would not vary with the level of the noise used to measure it. Unfortunately, this is not the case. Moore and Glasberg (1987) presented a summary of measurements of the auditory filter shape using maskers with notches placed asymmetrically about the signal frequency. They concluded that the lower skirt of the filter becomes less sharp with increasing level, while the higher skirt becomes slightly steeper. Glasberg and Moore (1990) re-analyzed the data from the studies summarized in that paper but used a modified fitting procedure including "corrections" for the transfer function of the middle ear. They also examined the data presented in Moore, Peters, and Glasberg (1990) and Shailer et al. (1990). The re-analysis led to the following conclusions:

(1) The auditory filter for a center frequency of 1 kHz is roughly symmetrical on a linear frequency scale when the level of the noise is approximately

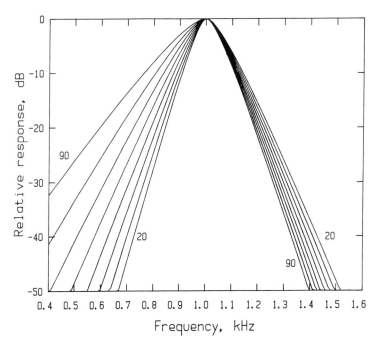

FIGURE 3.7. The shape of the auditory filter centered at 1 kHz, plotted for input sound levels ranging from 20 to 90 dB SPL/ERB. The relative response of the filter is plotted as a function of frequency. On the low-frequency side, the filter becomes progressively less sharply tuned with increasing sound level. On the high-frequency side, the sharpness of tuning increases slightly with increasing sound level. At moderate sound levels, the filter is approximately symmetrical on the linear frequency scale used. (Adapted from Moore and Glasberg 1987.)

51 dB/ERB. This corresponds to a noise spectrum level of about 30 dB. The auditory filters at other center frequencies are approximately symmetrical when the input levels to the filters are equivalent to the level of 51 dB/ERB at 1 kHz.
(2) The low-frequency skirt of the auditory filter becomes less sharp with increasing level.
(3) Changes in the slope of the high-frequency skirt of the filter with level are less consistent. At medium center frequencies (1 to 4 kHz), there is a trend for the slope to increase slightly with increasing level, but, at low center frequencies (Moore, Peters, and Glasberg 1990), there is no clear trend with level. The filters at high center frequencies (Shailer et al. 1990) show a slight decrease in the slope with increasing level.

The statements above are based on the assumption that, although the auditory filter is not linear, it may be considered as approximately linear at any given noise level. Furthermore, the sharpness of the filter is assumed to depend on the input level to the filter, not the output level. This issue is considered in Section 5.2. Figure 3.7 illustrates how the shape of the auditory filter varies with input level for a center frequency of 1 kHz.

5. Masking Patterns and Excitation Patterns

So far, masking experiments in which the frequency of the signal is held constant while the masker is varied have been discussed. These experiments are most appropriate for estimating the shape of the auditory filter at a given center frequency. However, many of the early experiments on masking did the opposite; the signal frequency was varied while the masker was held constant.

Wegel and Lane (1924) published the first systematic investigation of the masking of one pure tone by another. They determined the threshold of a signal with adjustable frequency in the presence of a masker with fixed frequency and intensity. The graph plotting masked threshold as a function of the frequency of the signal is known as a masking pattern, or sometimes as a masked audiogram. The results of Wegel and Lane were complicated by the occurrence of beats when the signal and masker were close together in frequency. To avoid this problem, later experimenters (e.g., Egan and Hake 1950) used a narrow band of noise as either the signal or the masker.

The masking patterns obtained in these experiments show steep slopes on the low-frequency side, of between 80 and 240 dB/octave for pure-tone masking and 55 to 190 dB/octave for narrowband noise masking. The slopes on the high-frequency side are less steep and depend on the level of the masker. A typical set of results is shown in Figure 3.8. Notice that, on the high-frequency side, the slopes of the curves tend to become shallower at high levels. Thus, if the level of a low-frequency masker is increased by, say, 10 dB,

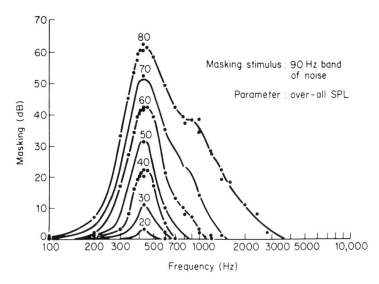

FIGURE 3.8. Masking patterns (masked audiograms) for a narrow band of noise centered at 410 Hz. Each curve shows the elevation in threshold of a sinusoidal signal as a function of signal frequency. The overall noise level for each curve is indicated in the figure. (Adapted from Egan and Hake 1950, by permission of the authors and Journal of the Acoustical Society of America.)

the masked threshold of a high-frequency signal is elevated by more than 10 dB; the amount of masking grows nonlinearly on the high-frequency side. This has been called the "upward spread of masking."

The masking patterns do not reflect the use of a single auditory filter. Rather, for each signal frequency, the listener uses a filter centered close to the signal frequency. Thus, the auditory filter is shifted as the signal frequency is altered. One way of interpreting the masking pattern is as a crude indicator of the excitation pattern of the masker. The excitation pattern of a sound is a representation of the activity or excitation evoked by that sound as a function of characteristic frequency or "place" in the auditory system (Zwicker and Feldtkeller 1967; Zwicker 1970). In the case of a masking pattern, it seems reasonable to assume that the signal is detected when the excitation it produces is some constant proportion of the excitation produced by the masker in the frequency region of the signal. Thus, the threshold of the signal as a function of frequency is proportional to the masker excitation level. The masking pattern should be parallel to the excitation pattern of the masker but shifted vertically by a small amount. In practice, the situation is not so straightforward, since the shape of the masking pattern is influenced by factors such as off-frequency listening and the detection of combination tones produced by the interaction of the signal and the masker (Greenwood 1971).

5.1 Relationship of the Auditory Filter to the Excitation Pattern

Moore and Glasberg (1983) have described a way of deriving the shapes of excitation patterns using the concept of the auditory filter. They suggested that the excitation pattern of a given sound can be thought of as the output of the auditory filters as a function of their center frequency. This idea is illustrated in Figure 3.9. The upper portion of the figure shows auditory filter shapes for five center frequencies. Each filter is symmetrical on the linear frequency scale used, but the bandwidths of the filters increase with increasing center frequency, as illustrated in Figure 3.6. The dashed line represents a 1-kHz sinusoidal signal whose excitation pattern is to be derived. The lower

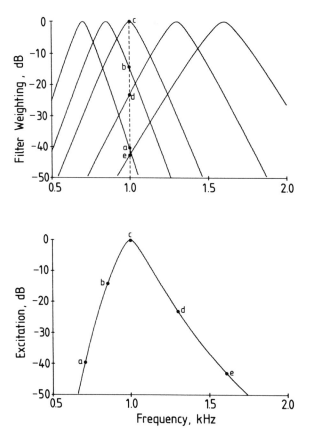

FIGURE 3.9. An illustration of how the excitation pattern of a 1-kHz sinusoid can be derived by calculating the outputs of the auditory filters as a function of their center frequency. The top half shows five auditory filters, centered at different frequencies, and the bottom half shows the calculated excitation pattern. (From Moore and Glasberg 1983.)

panel shows the output from each filter in response to the 1-kHz signal, plotted as a function of the center frequency of each filter; this is the desired excitation pattern.

To see how this pattern is derived, consider the output from the filter with the lowest center frequency. This has a relative output in response to the 1-kHz tone of about -40 dB, as indicated by point 'a' in the upper panel of Figure 3.9. In the lower panel of Figure 3.9, this gives rise to the point 'a' on the excitation pattern; the point has an ordinate value of -40 dB and is positioned on the abscissa at a frequency corresponding to the center frequency of the lowest filter illustrated. The relative outputs of the other filters are indicated, in order of increasing center frequency, by points 'b' to 'e,' and each leads to a corresponding point on the excitation pattern. The complete excitation pattern was actually derived by calculating the filter outputs for filters spaced at 10-Hz intervals. In deriving the excitation pattern, excitation levels were expressed relative to the level at the tip of the pattern, which was arbitrarily labeled as 0 dB. To calculate the excitation pattern for a 1-kHz tone with a level of, say, 60 dB, the level at the tip would be labeled as 60 dB, and all other excitation levels would correspondingly be increased by 60 dB.

Note that, although the auditory filters were assumed to be symmetrical on a linear frequency scale, the derived excitation pattern is asymmetrical. This happens because the bandwidth of the auditory filter increases with increasing center frequency. Note also that the excitation pattern has the same general form as the masking patterns shown in Figure 3.8.

5.2 Changes in Excitation Patterns with Level

One problem in calculating excitation patterns from filter shapes is how to deal with the level dependence of the auditory filter. In particular, it is necessary to know whether the shape of the auditory filter depends on the level of the input to the filter or on the level of the output of the filter. To examine this question, Moore and Glasberg (1987) calculated excitation patterns using both of these assumptions. An example of the results for a 1-kHz sinusoid at levels ranging from 20 to 90 dB SPL is shown in Figure 3.10.

The auditory filter shapes for each center frequency and masker level were calculated according to the assumptions outlined in Section 4.2. The high-frequency side of each excitation pattern is determined by the low-frequency branches of the auditory filters (i.e., by the values of p_l), and vice versa. For the left-hand panels in Figure 3.10, the output level of each filter was assumed to determine its shape. For the right-hand panels, the input level was assumed to determine the shape.

As described in Section 5, the shapes of excitation patterns for narrowband stimuli as a function of level can be determined approximately from their masking patterns (Zwicker and Feldtkeller 1967). The patterns shown in the right-hand panels of Figure 3.10 closely resemble masked audiograms at similar masker levels, whereas those in the left-hand panels are very different

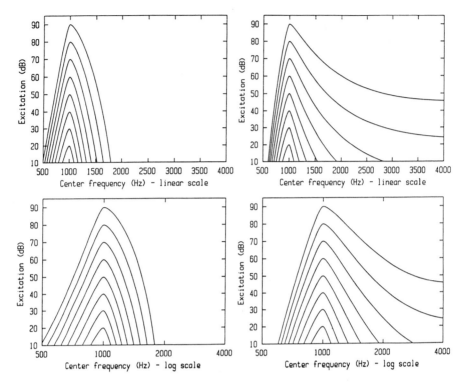

FIGURE 3.10. Excitation patterns calculated according to the procedure described in the text for 1-kHz sinusoids ranging in level from 20 to 90 dB SPL in 10-dB steps. The frequency scale is linear for the upper panels and logarithmic for the lower panels. The left panels show excitation patterns calculated on the assumption that the level at the output of the auditory filter is the variable determining its shape. The right panels show excitation patterns calculated assuming that the level at the input to the filter determines its shape. (From Moore and Glasberg 1987.)

in form and do not show the classic "upward spread of masking." It may be concluded that the critical variable determining the auditory filter shape is the input level to the filter. A similar conclusion was reached by Lutfi and Patterson (1984).

Unfortunately, the situation cannot be as simple as this. For example, it is unlikely that a 16-kHz tone at 90 dB SPL would affect the auditory filter shape at a center frequency of 1 kHz. In general, for sounds with complex broadband spectra, it seems likely that only components which produce a significant output from a given auditory filter have any influence in determining the shape of that filter. A plausible extension to the simple rule given is that the shape of the auditory filter is determined primarily by the input level of the component that produces the greatest output from the filter. When the input spectrum is continuous, or contains closely spaced components, the power of the components is summed within a range of one ERB (around the frequencies of the components in question) to determine the effective input level.

6. Phenomena Reflecting the Influence of Auditory Filtering

6.1 The Loudness of Complex Sounds

Consider a complex sound of fixed energy (or intensity) having a bandwidth of W. If W is less than a certain bandwidth, called the CB for loudness, then the loudness of the sound is more or less independent of W; the sound is judged to be about as loud as a pure tone of equal intensity lying at the center frequency of the band. However, if W is increased beyond the CB for loudness, the loudness of the complex begins to increase. This has been found to be the case for bands of noise (Zwicker, Flottorp, and Stevens 1957) and for complexes consisting of pure tones whose frequency separation is varied (Scharf 1961, 1970). An example for noise bands is illustrated in Figure 3.11.

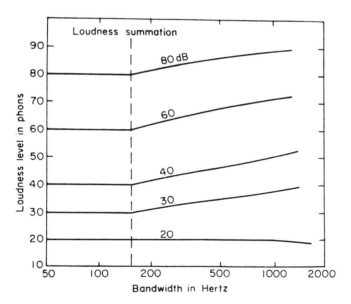

FIGURE 3.11. The loudness level in phons of a band of noise centered at 1 kHz, measured as a function of the width of the band. For each of the curves, the overall sound level was constant, and its value, in dB SPL, is indicated in the figure. The dashed line shows that the bandwidth at which loudness begins to increase is roughly the same at all levels tested (except that no increase occurs at the lowest level). (Adapted from Zwicker and Feldtkeller 1967.)

The CB for loudness for the data in Figure 3.11 is about 160 Hz for a center frequency of 1000 Hz.

Moore and Glasberg (1986) showed that the variation of loudness with bandwidth can be accounted for rather well by a modification of a model for predicting loudness proposed by Zwicker and Scharf (1965). The model involves two stages. In the first, the excitation pattern of the sound of interest is calculated. Rather than inferring excitation patterns from masking patterns, as was done by Zwicker and Scharf, Moore and Glasberg calculated

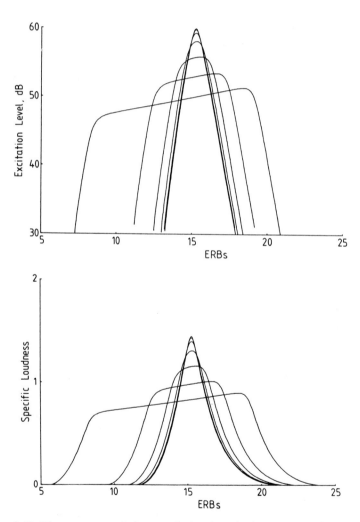

FIGURE 3.12. The upper panel shows calculated excitation patterns for a series of sounds with fixed overall level, but variable bandwidth, centered at 1 kHz. The lower panel shows specific loudness patterns corresponding to these excitation patterns. (From Moore and Glasberg 1986.)

excitation patterns according to the procedure described by Moore and Glasberg (1983; see Section 5.1). The results of this are illustrated in the upper panel of Figure 3.12, which shows excitation patterns for a sinusoid and for complex sounds of various bandwidths (20, 60, 140, 220, 300, 620, and 1260 Hz), all centered at 1 kHz. The frequency axis is scaled in units of the ERB of the auditory filter (ERB scale; see Moore and Glasberg 1983; Glasberg and Moore 1990; an equation relating number of ERBs to frequency is: number of ERBs, $E = 21.4 \log_{10}(4.37F + 1)$, where F is frequency in kHz).

In the second stage of the model, the excitation level at each frequency is converted into specific loudness via a power-law relationship. The lower panel in Figure 3.12 shows the specific loudness patterns corresponding to the excitation patterns in the top panel. The loudness of each sound corresponds to the total area under its specific loudness pattern.

The reason that loudness remains constant at first as bandwidth increases is that the decrease in specific loudness at the tip of the pattern is almost exactly compensated by the increase on the skirts of the pattern. However, beyond a certain bandwidth (the CB for loudness), the increase on the skirts is more than enough to offset the decrease at the tip, and loudness then increases. It should be noted that the bandwidth at which loudness starts to increase is not numerically equal to the ERB of the auditory filter at that center frequency. The patterns in Figure 3.12 were calculated assuming the ERB of the auditory filter was 128 Hz at 1 kHz. The bandwidth at which loudness is predicted to increase is about 165 Hz, in excellent agreement with the empirically obtained value.

At low overall levels, loudness hardly changes with bandwidth. This can be explained by the model because, close to threshold, the specific loudness changes rapidly with excitation level. The increasing spread of the specific loudness pattern with increasing bandwidth is almost exactly offset by the decrease in specific loudness in the main part of the pattern, so that the total area remains roughly constant.

6.2 *The Threshold of Complex Sounds*

When two tones with a small frequency separation are presented together, a sound may be heard even when either tone by itself is below threshold. Gässler (1954) measured the threshold of multitone complexes consisting of evenly spaced sinusoids. The tones were presented both in quiet and in a special background noise, chosen to give the same masked threshold for each component in the signal. As the number of tones in a complex was increased, the threshold, specified in terms of total energy, remained constant until the overall spacing of the tones reached a certain bandwidth, the CB for threshold. Thereafter, the threshold increased by about 3 dB per doubling of bandwidth. The CB for a center frequency of 1 kHz was estimated to be about 180 Hz. These results were interpreted as indicating that the energies of the individual components in a complex sound will sum in the detection of that sound, provided the components lie within a critical band. When the components are distributed over more than one critical band, detection is based on the single band giving the highest detectability.

Other data are not in complete agreement with those of Gässler. For example, Spiegel (1981) measured the threshold for a noise signal of variable bandwidth centered at 1 kHz in a broadband background noise masker. The threshold for the signal as a function of bandwidth did not show a breakpoint corresponding to the CB, but increased monotonically as the bandwidth increased beyond 50 Hz. The slope beyond the CB was close to 1.5 dB per doubling of bandwidth. Higgins and Turner (1990) have suggested that the discrepancy may be explained by the fact that Gässler widened the bandwidth, keeping the upper edge of the complex fixed in frequency, while Spiegel used stimuli with a fixed center frequency. However, other results clearly show that the ear is capable of combining information over bandwidths much greater than the CB. For example, Buus et al. (1986) showed that multiple, widely spaced sinusoidal components were more detectable than any of the individual components.

The results of Spiegel (1981) and Buus et al. (1986) should not be interpreted as evidence against the concept of the auditory filter. They do indicate, however, that detection of complex signals is not based on the output of a single auditory filter. Rather, information can be combined across filters to improve performance.

6.3 Sensitivity to Relative Phase

Under some conditions, an amplitude-modulated (AM) sine wave and a frequency-modulated (FM) sine wave may have components (corresponding to the carrier frequency and two "side bands") which are identical in frequency and amplitude, the only difference between them being in the relative phase of the components (an FM wave actually contains many components but, for small modulation depths, only the first two side bands are important). If, then, the two types of wave are perceived differently, the difference is likely to arise from a sensitivity to the relative phase of the components.

Zwicker (1952) and Schorer (1986) have measured one aspect of the perception of such stimuli, namely, the just detectable amounts of amplitude or frequency modulation for various rates of modulation. They found that, for high rates of modulation where the frequency components were widely spaced, the detectability of FM and AM was equal when the components in each type of wave were of equal amplitude; the threshold appeared to be determined by the ability to detect the side bands. However, when all three components fell within a CB, AM could be detected when the relative levels of the side bands were lower than for a wave with a just-detectable amount of FM. Thus, it appears that changes in the relative phase of the components only affect modulation detection when those components lie within a CB.

This does not seem to apply to the perception of suprathreshold levels of modulation or to other aspects of phase sensitivity. For example, subjects can detect phase changes between the components in complex sounds in which the components are separated by considerably more than a CB (Patterson 1987). Also, subjects can compare the relative phase of modulation of carriers in widely separated frequency regions (see Yost and Sheft, Chapter 6).

6.4 The Audibility of Partials in Complex Tones

According to Ohm's (1843) Acoustical Law, the ear is able to hear pitches corresponding to the individual sinusoidal components in a complex periodic sound. In other words, humans can "hear out" the individual partials. Plomp (1964) and Plomp and Mimpen (1968) used a complex tone with 12 sinusoidal components to investigate the limits of this ability. The listener was presented with two comparison tones, one of which had the same frequency as a partial in the complex while the other lay halfway between that frequency and the frequency of the adjacent higher or lower partial. The listener had to judge which of these two tones was a component of the complex. Two types of complex were used: a harmonic complex containing harmonics 1 to 12, where the frequencies of the components were integral

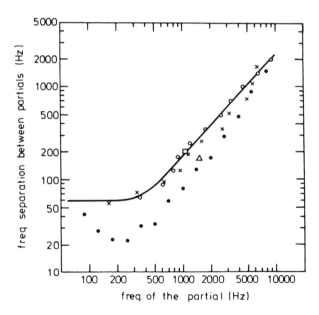

FIGURE 3.13. The frequency separation between adjacent partials required for a given partial to be "out" from a complex tone with either two equal-amplitude components (solid circles) or many equal-amplitude components (open symbols and crosses), plotted as a function of the frequency of the partial. (From Plomp 1976, by permission of the author and publisher.)

multiples of that of the fundamental; and a nonharmonic complex, where the frequencies of the components were mistuned from simple frequency ratios. For both kinds of complex, only the first five to eight components could be "heard out." These results are shown by the open circles (harmonic complex) and crosses (inharmonic complex) in Figure 3.13.

If it is assumed that a partial will only be distinguished when it is separated from its neighbor by at least one CB, then the results can be used to estimate the CB. At low frequencies, the CBs estimated in this way are smaller than the traditional values of the CB (Zwicker 1961). However, the results are in

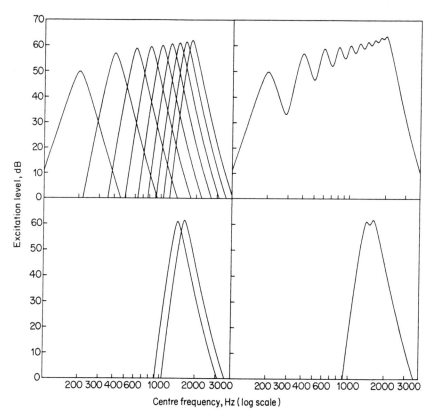

FIGURE 3.14. The top left panel shows the excitation pattern for each harmonic in a complex tone containing the first nine harmonics of a 200-Hz fundamental. The top right panel shows the excitation pattern resulting from adding the nine harmonics together. The bottom left panel shows the excitation patterns for the seventh and eighth harmonics separately, and the bottom right panel shows the excitation pattern resulting from adding those two harmonics. For the two-tone complex, neurons with characteristic frequencies (CFs) below 1400 Hz would be phase locked primarily to the lower of the two harmonics, while neurons with CFs above 1600 Hz would be phase locked primarily to the upper of the two harmonics. (From Moore, 1989.)

quite good agreement with the values of the ERB given by Equation 6 (see Section 4.1).

When Plomp (1964) repeated the experiment using a two-tone complex, he found that the partials could be distinguished at smaller frequency separations than were found for multitone complexes (solid circles in Figure 3.13). The discrepancy was particularly large at low center frequencies. A possible explanation for the discrepancy comes from the idea that "hearing out" partials from complex tones depends partly on the analysis of temporal patterns of firing in the auditory nerve. Figure 3.14 shows excitation patterns for a two-tone complex and a multitone complex. For a multitone complex, the excitation at center frequencies corresponding to the higher partials arises from the interaction of several partials of comparable effectiveness. Thus, there is no center frequency where the temporal pattern of response is determined primarily by one partial. However, for a two-tone complex, there are certain center frequencies where the temporal pattern of response is dominated by one or the other of the component tones; this occurs at center frequencies just below and above the frequencies of the two tones. Thus, the temporal patterns of firing in neurons tuned to these frequencies could signal the individual pitches of the component tones. This could explain why partials are more easily "heard out" from a two-tone complex than a multitone complex.

7. The Origin of the Auditory Filter

The physiological basis of the auditory filter is still uncertain, although the frequency-resolving power of the basilar membrane is almost certainly involved. Indeed, there are many similarities between the frequency selectivity measured on the basilar membrane and frequency selectivity measured psychophysically (Moore 1986). It appears that the CB, or the ERB of the auditory filter, corresponds to a constant distance along the basilar membrane (Greenwood 1961, 1990; Moore 1986); in humans, each ERB corresponds to about 0.9 mm, regardless of the center frequency. Further, the ERB of the auditory filter measured behaviorally in animals corresponds rather well with the ERB of tuning curves measured in single neurons of the auditory nerve in the same species. This is illustrated in Figure 3.15 (data from Evans, Pratt, and Cooper 1989). Behavioral ERBs were measured using bandstop noise (BSN), as described in Section 3.2, and comb-filtered noise (CFN), as described in Section 3.3.

Although this good correspondence between behavioral and neural ERBs suggests that the frequency selectivity of the auditory system is largely determined in the cochlea, it is possible that there is a sharpening process following the basilar membrane which results in an enhancement of frequency selectivity. This might be achieved by a process of lateral suppression or lateral inhibition. In such a process, weak inputs to one group of receptors

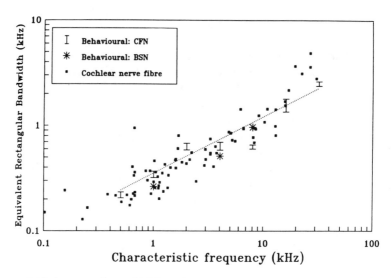

FIGURE 3.15. A comparison of ERBs estimated from behavioral masking experiments and from neurophysiological measurements of the tuning curves of single neurons in the auditory nerve. All data were obtained from guinea pigs. There is a good correspondence between behavioral and neural data. (From Evans, Pratt, and Cooper 1989, by permission of the authors.)

(hair cells or neurons) would be suppressed by stronger inputs to adjacent receptors. This would result in a sharpening of the excitation pattern evoked by a stimulus.

8. Time Course of Development of the Auditory Filter

If the critical band mechanism depended on a neural inhibitory process, then it would be expected that the process would take several milliseconds to become effective. Thus, the CB measured with very brief signals should be greater than that measured with longer signals. The data relevant to this issue are not entirely clear-cut. Port (1963) found that measures of loudness as a function of bandwidth for short duration stimuli (as short as 1 ms) gave the same values for the CB as at long durations (several hundred ms). Some other experimental results (e.g., Zwicker 1965a,b; Srinivasan 1971) have been taken to indicate a wider CB at short durations. At present, the reasons for the discrepancies are not entirely clear. Zwicker and Fastl (1972), in a review of this area, concluded that the results which have been taken to indicate a wider CB at short durations can, in fact, be explained in other ways, so that it is not necessary to assume a development of the critical band with time. Rather, the critical band mechanism may be viewed as a system of filters which are permanently and instantaneously present. This view has been supported by an experiment of Moore et al. (1987). They estimated the shape of the auditory filter using the notched-noise method. The signal was a brief (20 ms) tone presented at the start, the temporal center, or the end of the 400-ms masker. The auditory filter shapes derived from the results did not change significantly with signal delay, suggesting that the selectivity of the auditory filter does not develop over time. On the other hand, measures of frequency selectivity obtained with tonal maskers do show a development of frequency selectivity with time (Bacon and Viemeister 1985; Bacon and Moore 1986).

Overall, these results make it unlikely that any neural inhibitory process is involved in the formation of the critical band. It is possible, however, that some form of lateral suppression might operate at a very early stage in the processing of the auditory stimulus, for example, at the level of the hair cell (Sellick and Russell 1979). Such a process could operate very quickly.

9. Evidence for Lateral Suppression from Nonsimultaneous Masking

The results from experiments on simultaneous masking can generally be explained quite well on the assumption that the auditory filters are approximately linear. However, measurements from single neurons in the auditory nerve show significant nonlinearities. In particular, the response to a tone of a given frequency can sometimes be suppressed by a tone with a different frequency, giving the phenomenon known as two-tone suppression (Arthur, Pfeiffer, and Suga 1971). For other complex signals, similar phenomena occur and are given the general name lateral suppression. This can be characterized in the following way. Strong activity at a given characteristic frequency (CF) can suppress weaker activity at adjacent CFs. In this way, peaks in the excitation pattern are enhanced relative to adjacent dips. The question now arises as to why the effects of lateral suppression are not usually seen in experiments on simultaneous masking.

Houtgast (1972) has argued that simultaneous masking is not an appropriate tool for detecting the effects of lateral suppression. He argued that, in simultaneous masking, the masking stimulus and the test tone are processed simultaneously in the same channel. Thus, any suppression in that channel will affect the neural activity caused by both the test tone and the masking noise. In other words, the signal-to-noise ratio in a given frequency region will be unaffected by lateral suppression and, thus, the threshold of the test tone will remain unaltered.

Houtgast suggested that this difficulty could be overcome by presenting the masker and the test tone successively (i.e., by using a forward masking technique). If lateral suppression does occur, then its effects will be seen in forward masking provided that: (1) in the chain of levels of neural processing, the level at which the lateral suppression occurs is not later than the level at which most of the forward masking effect arises; and (2) the suppression built up by the masker has decayed by the time the test tone is presented (otherwise the problems described for simultaneous masking will be encountered).

Houtgast (1972) used a repeated-gap masking technique, in which the masker was presented with a continuous rhythm of 150 ms on, 50 ms off. Probe tones with a duration of 20 ms were presented in the gaps. In one experiment, he used as maskers high-pass and low-pass noises with sharp spectral cutoffs (96 dB/octave). He anticipated that lateral suppression would result in an enhancement of the neural representation of the spectral edges of the noise. Neurons with CFs well within the passband of the noise should have their responses suppressed by the activity at adjacent CFs. However, for neurons with a CF corresponding to a spectral edge in the noise, there should be a release from suppression, owing to the low activity in neurons with CFs outside the spectral range of the noise. This should be revealed as an increase in the threshold of the probe when its frequency coincided with the spectral edge of each noise band. The results showed the expected edge effects while no such effects were found in simultaneous masking. Thus, nonsimultaneous masking does reveal the type of effects which a suppression process would produce.

Houtgast (1972) also noted that when the bursts of probe tone are just above the threshold, they sound like a continuous tone. Only at higher levels of the probe tone is the perception in accord with the physical time pattern, namely, a series of tone bursts. Houtgast called the level of the probe tone at which its character changed from pulsating to continuous the pulsation threshold. He suggested the following interpretation of the phenomenon "when a tone and a stimulus, S, are alternated (alternation cycle about 4 Hz), the tone is perceived as being continuous when the transition from S to tone causes no (perceptible) increase of nervous activity in any frequency region" (Houtgast 1972). In terms of patterns of excitation, this suggests that "the peak of the nervous activity pattern of the tone at the pulsation threshold level just reaches the nervous activity pattern of S." Given this hypothesis, the pulsation threshold for a test tone as a function of frequency can be considered to map out the excitation pattern of the stimulus S (including the effects of suppression). For convenience, the measurement of the pulsation threshold will be referred to as a nonsimultaneous masking technique.

Following the pioneering work of Houtgast, many workers (for a review see Moore and O'Loughlin 1986) have reported that there are systematic differences between the results obtained using simultaneous and nonsimultaneous masking techniques. One major difference is that nonsimultaneous masking reveals effects that can be directly attributed to suppression. A good demonstration of this involves a psychophysical analogue of neural two-tone suppression. Houtgast (1973, 1974) measured the pulsation threshold for a 1-kHz "signal" alternating with a 1-kHz "masker." He then added a second tone to the "masker" and measured the pulsation threshold again. He found that sometimes the addition of this second tone produced a reduction in the pulsation threshold and attributed this to a suppression of the 1-kHz component in the "masker" by the second component. If the 1-kHz component is suppressed, then there will be less activity in the frequency region around 1 kHz, producing a drop in the pulsation threshold. The second tone was most effective as a "suppressor" when it was somewhat more intense than the 1-kHz component and above it in frequency. A "suppression" of about 20 dB could be produced by a "suppressor" at 1.2 kHz. Houtgast (1974) mapped out the combinations of frequency and intensity over which the "suppressor" produced a "suppression" exceeding 3 dB. He found two regions, one above 1 kHz and one below, as illustrated in Figure 3.16. The regions found were similar to the suppression areas which can be observed in single neurons of the auditory nerve. Similar results have been found using a forward masking technique (Shannon 1976).

Under some circumstances, the reduction in threshold (unmasking) produced by adding one or more extra components to a masker can be partly

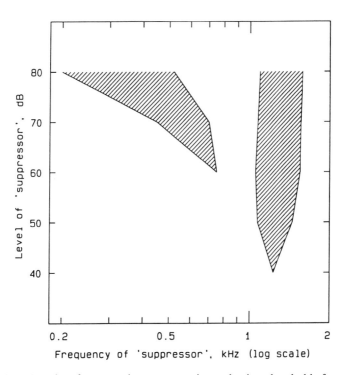

FIGURE 3.16. Results of an experiment measuring pulsation thresholds for a 1-kHz "signal" alternated with a two component "masker." One component was a 1-kHz tone which was fixed in level at 40 dB. The second component (the "suppressor") was a tone which was varied both in frequency and in level. The shaded areas indicate combinations of frequency and level where the second tone reduced the pulsation threshold by 3 dB or more. (Adapted from Houtgast 1974.)

explained in terms of additional cues provided by the added components, rather than in terms of suppression. Specifically, in forward masking, the added components may reduce "confusion" of the signal with the masker by indicating exactly when the masker ends and the signal begins (Moore 1980, 1981; Moore and Glasberg 1982a, 1985; Neff 1985; Moore and O'Loughlin 1986). This may have led some researchers to overestimate the magnitude of suppression, as indicated in nonsimultaneous masking experiments. However, it seems clear that not all unmasking can be explained in this way.

10. The Enhancement of Frequency Selectivity Revealed in Nonsimultaneous Masking

A second major difference between simultaneous and nonsimultaneous masking is that the frequency selectivity revealed in nonsimultaneous masking is greater than that revealed in simultaneous masking. A well-studied example of this is the psychophysical tuning curve (PTC). PTCs determined in forward masking, or using the pulsation threshold method, are typically sharper than those obtained in simultaneous masking (Moore 1978). An example is given in Figure 3.17. The difference is particularly marked on the high-frequency side of the tuning curve. According to Houtgast (1974), this difference arises because the internal representation of the masker (its excitation pattern) is sharpened by a suppression process, with the greatest sharpening occurring on the low-frequency side. In simultaneous masking, the effects of suppression are not seen since any reduction of the masker activity in the frequency region of the signal is accompanied by a similar reduction in signal-evoked activity. In other words, the signal-to-masker ratio in the frequency region of the signal is unaffected by the suppression. In forward masking, on the other hand, the suppression does not affect the signal. For maskers with frequencies above that of the signal, the effect of suppression is to sharpen the excitation pattern of the masker, resulting in an increase of the masker level required to mask the signal. Thus, the suppression is revealed as an increase in the slopes of the PTC. This explanation of the difference between PTCs in simultaneous and forward masking will be referred to as the SPEX model since, in both simultaneous and forward masking, the masking is produced by the **sp**read of masker **ex**citation to the frequency region of the signal.

Although this explanation is appealing, it is not the only way of accounting for the sharper PTC obtained in nonsimultaneous masking. An alternative explanation is that, in simultaneous masking, the low-level signal may be suppressed by the masker so that it falls below absolute threshold. The neural data indicate that tones falling outside of the region bounded by the neural tuning curve can produce suppression. Thus, the PTC in simultaneous masking might map out the boundaries of the more broadly tuned suppression region (Delgutte 1988). According to this model, masking by simultaneous maskers remote from the signal frequency is produced by suppression of the

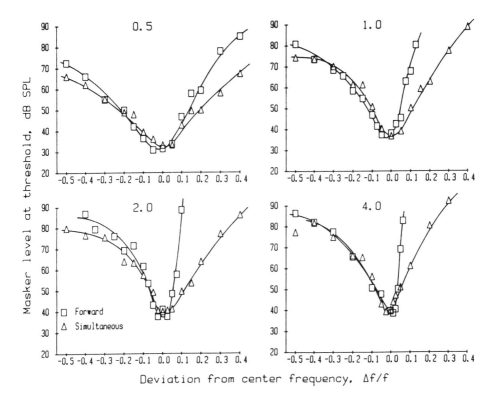

FIGURE 3.17. Comparison of psychophysical tuning curves (PTCs) determined in simultaneous masking (triangles) and forward masking (squares). The masker frequency is plotted as deviation from the center frequency divided by the center frequency ($\Delta f/f$). The center frequency is indicated in kHz in each panel. A low-level notched noise was gated with the masker to provide a consistent detection cue in forward masking and to restrict off-frequency listening. (From Moore, Glasberg, and Roberts 1984.)

signal to absolute threshold, rather than by spread of excitation. Thus, this will be referred to as the STAT model. The signal in nonsimultaneous masking could not be suppressed in this way, so, in this case, the masking would be produced by spread of excitation.

A major difference between these two explanations is that the first assumes that suppression will sharpen the excitation pattern of a single sinusoidal component, whereas the second assumes only that one component may suppress another; suppression will not necessarily affect the internal representation of a single component.

It remains unclear which of these two explanations is correct. Moore and Glasberg (1982b) concluded, on the basis of a psychophysical experiment, that the SPEX model was correct, and a physiological experiment by Pickles

(1984) supported this view. However, Delgutte (1990) has presented physiological evidence suggesting that simultaneous masking by intense low-frequency tones (upward spread of masking) is largely due to suppression rather than spread of excitation, supporting the STAT model.

Several other methods of estimating frequency selectivity have indicated sharper tuning in nonsimultaneous masking than in simultaneous masking. For example, auditory filter shapes estimated in forward masking using a notched-noise masker have smaller bandwidths and greater slopes than those estimated in simultaneous masking (Moore and Glasberg 1981; Moore et al. 1987). Similarly, auditory filter shapes estimated with the pulsation threshold method (a kind of nonsimultaneous masking) using rippled noise are sharper than those measured in simultaneous masking (Houtgast 1977). This encourages the belief that a general consequence of suppression is an enhancement of frequency selectivity.

It may be concluded that results from nonsimultaneous masking do clearly show the types of effects which would be expected if a suppression mechanism was operating. The level at which the effects occur is uncertain, but the most common assumption has been that they arise at a very early stage in auditory processing, possibly at or close to the point of transduction of mechanical movements to neural impulses. The effects of suppression may be characterized by stating that strong excitation at a given CF can reduce the excitation at adjacent CFs. It seems likely that the effects of suppression are not (usually) revealed in simultaneous masking.

11. Introduction to Pitch Perception

Pitch may be defined as "that attribute of auditory sensation in terms of which sounds may be ordered on a scale extending from high to low" (American National Standards Institute 1973). In other words, variations in pitch give rise to a sense of melody. Pitch is related to the repetition rate of the waveform of a sound; for a pure tone, this corresponds to the frequency and for a periodic complex tone to the fundamental frequency. There are, however, exceptions to this simple rule (see Section 16.1). Since pitch is a subjective attribute, it cannot be measured directly. Assigning a pitch value to a sound is generally understood to mean specifying the frequency of a pure tone having the same subjective pitch as the sound.

12. Theories of Pitch Perception for Pure Tones

For many years, there have been two different theories of pitch perception (see Yost and Sheft, Chapter 6). One, the "place" theory, has two distinct postulates. The first is that the stimulus undergoes a spectral analysis in the inner ear, so that different frequencies (or frequency components in a complex stimulus) excite different places along the basilar membrane and, hence, neurons with different CFs. The second is that the pitch of a stimulus is related to the pattern of excitation produced by that stimulus; for a pure tone, the pitch is generally assumed to correspond to the position of maximum excitation. The first of these two postulates is now well established and has been confirmed in a number of independent ways, including direct observation of the movement of the basilar membrane. The second is still a matter of dispute.

An alternative to the place theory, called the "temporal" theory, suggests that the pitch of a stimulus is related to the time pattern of the neural impulses evoked by that stimulus. Nerve firings tend to occur at a particular phase of the stimulating waveform (phase locking), and, thus, the intervals between successive neural impulses approximate integral multiples of the period of the stimulating waveform. The temporal theory could not work at very high frequencies, since phase locking does not occur for frequencies above about 5 kHz. However, the tones produced by musical instruments, the human voice, and most everyday sound sources all have fundamental frequencies below this range.

13. The Perception of the Pitch of Pure Tones

13.1 The Frequency Discrimination of Pure Tones

There have been two common ways of measuring frequency discrimination. One measure, called the DLF (difference limen for frequency), involves the discrimination of successive steady tones with slightly different frequencies. A second measure, called the FMDL (frequency modulation difference limen), uses tones that are frequency modulated (FM) at a low rate (typically 2 to 4 Hz). The amount of modulation required for detection of the modulation is determined.

Early studies of frequency discrimination mainly measured FMDLs (e.g., Shower and Biddulph 1931; Zwicker 1956). Recent studies have concentrated more on the measurement of DLFs. A summary of the results of some of these studies is given in Figure 3.18, taken from Wier, Jesteadt, and Green (1977). Expressed in Hz, the DLF is smallest at low frequencies and increases monotonically with increasing frequency. Expressed as a proportion of center frequency, the DLF tends to be smallest for middle frequencies and larger for very high and very low frequencies. Wier, Jesteadt, and Green (1977) found that the data describing the DLF as a function of frequency fell on a straight line when plotted as log(DLF) against $\sqrt{}$(frequency); the axes are scaled in this way in Figure 3.18. This probably has no particular theoretical significance. FMDLs, shown as the dashed line in Figure 3.18, tend to vary less with frequency than DLFs. Both DLFs and FMDLs tend to get somewhat smaller as the sound level increases.

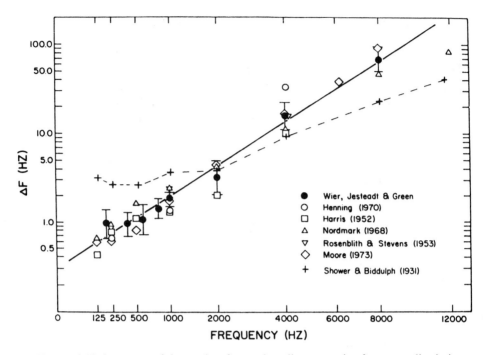

FIGURE 3.18. Summary of the results of several studies measuring frequency discrimination thresholds. The thresholds, ΔF, are plotted in Hz as a function of frequency (on a square-root scale). All of the studies except that of Shower and Biddulph (1931) measured DLFs; they measured FMDLs. (From Wier, Jesteadt, and Green 1977, by permission of the authors and Journal of the Acoustical Society of America.)

Zwicker (1970) has attempted to account for frequency discrimination in terms of changes in the excitation pattern evoked by the stimulus when the frequency is altered. His model is generally regarded as a place model, even though the excitation level might be coded partly in the timing of neural impulses. Zwicker inferred the shapes of the excitation patterns from masked audiograms such as those shown in Figure 3.8. In his original formulation of the model, Zwicker intended it to apply only to FMDLs; others (e.g. Moore, 1973a) have tried to apply the model to account for DLFs.

According to Zwicker's (1970) model, a change in frequency will be detected whenever the excitation level at some point on the pattern changes by more than a certain threshold value. Zwicker suggested that this value was about 1 dB. The change in excitation level is always greatest on the steeply sloping low-frequency side of the excitation pattern (see Figure 3.10). Thus, in this model, the detection of a change in frequency is functionally equivalent to the detection of a change in level on the low-frequency side of the excitation pattern. Maiwald (1967) has shown that the steepness of the low-frequency side is roughly constant when expressed in units of the CB rather than in terms of linear frequency. The slope has a value of 27 dB/bark (a bark is a unit of one CB). Thus, Zwicker's (1970) model predicts that the frequency DL at any given frequency should be a constant fraction (1/27) of the CB at that frequency; a change in frequency of 1/27 bark should give a change in excitation level of 1 dB. FMDLs do conform fairly well to the predictions of the model. However, DLFs vary more with frequency than predicted by the model; at low frequencies, DLFs are smaller than predicted while at high frequencies they are slightly larger than predicted (Moore and Glasberg 1986).

Henning (1966) has pointed out a problem with the measurement of frequency difference limens at high frequencies: the frequency changes may be accompanied by correlated loudness changes. These occur because the response of headphones on real ears is generally very irregular at high frequencies, and there is also a loss of absolute sensitivity of the observer at high frequencies. These loudness changes may provide observers with usable cues in the detection of frequency changes. To prevent subjects from using these cues, Henning (1966) measured DLFs for tones whose level was varied randomly from one stimulus to the next. The random variation in level produced changes in loudness which were large compared with those produced by the frequency changes. Henning found that DLFs at high frequencies (4 kHz and above) were markedly increased by the random variation in level, whereas those at low frequencies were not.

Although Henning's (1966) experiment was carried out primarily to assess the importance of loudness cues in frequency discrimination, his data also provided a test of Zwicker's (1970) model. If the detection of frequency changes was based on the detection of changes in excitation level on the low-frequency side of the excitation pattern, then the introduction of random variations in level should have markedly impaired frequency discrimination; the small change in excitation level produced by a given frequency change would now be superimposed on much larger random changes in level. This predicted impairment of frequency discrimination was found only at high frequencies. Thus, Henning's data were consistent with Zwicker's model for high frequencies but not for low frequencies. However, Emmerich, Ellermeier, and Butensky (1989) repeated Henning's experiment and found that random variations in level did result in larger DLFs at low frequencies (0.5 to 4.0 kHz). Although these data appear to support Zwicker's model, Emmerich and co-workers argued that this was not necessarily true. They presented evidence that the impairment produced by the randomization of level could be attributed to slight changes in the pitches of the sinusoidal signals with changes in level (see below for a description of such changes).

Moore and Glasberg (1989) also measured DLFs for tones whose level was randomized from one stimulus to the next, but they randomized the level over a relatively small range (6 dB instead of 20 dB used by Henning [1966] and Emmerich, Ellermeier, and Butensky [1989]) to minimize changes of pitch with level. The 6-dB range was still large relative to the changes in excitation level produced by small frequency changes. Moore and Glasberg found only a very small effect of the randomization of level; the DLFs were, on average, only 15 % greater than those measured with the level fixed. The DLFs measured with the level randomized were smaller than predicted by Zwicker's model for frequencies from 0.5 to 4.0 kHz. At 6.5 kHz (the highest frequency tested), the data were consistent with the model.

An alternative way of testing Zwicker's model is to use a stimulus which has energy distributed over a range of frequencies and whose magnitude spectrum has a certain slope. The slope of the excitation pattern of such a stimulus cannot be steeper than the slope of the physical spectrum. This approach was adopted by Moore (1972, 1973a) who measured DLFs for tone pulses of various durations. Below some critical duration, the slope of the spectral envelope will be less than the slope of the excitation pattern evoked by a long-duration pure tone. Thus, if Zwicker's model is correct, this physical slope will limit performance at short durations. The results showed that, at short durations, observers did better than predicted for all frequencies up to 5 kHz. At about this frequency, the DLs showed a sharp increase in value (see Figure 3.19).

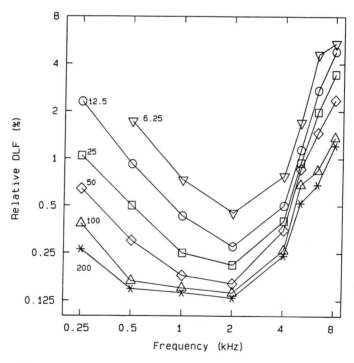

FIGURE 3.19. Values of the DLF plotted as a function of center frequency and expressed as a percentage of center frequency (log scale). The number by each curve is the duration of the tone pulses in ms. Notice the sharp increase in the size of the DLFs which occurs around 5 kHz. (Data from Moore 1973a.)

In summary, Zwicker's (1970) model predicts that frequency DLs should vary with frequency in the same way as the CB. It also predicts that random variations in level should markedly increase frequency DLs. The results for DLFs are not consistent with these predictions. DLFs vary more with frequency than the CB, and the effect of randomizing level is smaller than predicted, except at high frequencies. Furthermore, DLFs for short-duration tones are smaller than predicted by Zwicker's model, except above 5 kHz. Other data, reviewed in Moore and Glasberg (1986), also suggest that Zwicker's model does not adequately account for DLFs. The results for FMDLs reviewed here are generally consistent with Zwicker's model, although other data suggest that it may not be entirely adequate even for these (Feth 1972; Coninx 1978).

Although the concentration has been on Zwicker's (1970) model, other place models (e.g., Henning 1967; Siebert 1968, 1970) also have difficulty in accounting for DLFs. All place models predict that frequency discrimination should be related to frequency selectivity; the sharper the tuning of peripheral filtering mechanisms, the smaller should be the frequency DL. Thus, all place models predict that the frequency DL should vary with frequency in the same way as the CB. The failure of this prediction suggests that some other mechanism is involved.

The results are consistent with the idea that DLFs are determined by temporal information (phase locking) for frequencies up to about 4 to 5 kHz, and by place information above that. The precision of phase locking decreases with increasing frequency above 1 to 2 kHz, and it is completely absent above about 5 kHz. This can explain why DLFs increase markedly above this frequency. Goldstein and Srulovicz (1977) have shown that it is possible to predict the dependence of the frequency DL on frequency and duration by assuming that the auditory system processes the time intervals between successive nerve impulses, ignoring higher order temporal dependencies.

The two mechanisms under discussion are not, of course, mutually exclusive; it is likely that place information is available over almost all of the auditory range but that temporal mechanisms allow better frequency discrimination for frequencies up to about 4 to 5 kHz.

13.2 *The Perception of Musical Intervals*

Two tones which are separated in frequency by an interval of one octave (i.e., one has twice the frequency of the other) sound similar and are judged to have the same name on the musical scale (for example, C or D). This has led several theorists to suggest that there are at least two dimensions to musical pitch. One aspect is related monotonically to frequency (for a pure tone) and is known as "tone height." The other is related to pitch class (i.e., name note) and is called "tone chroma" (Bachem 1950; Shepard 1964).

If subjects are presented with a pure tone of a given frequency, f_1, and are asked to adjust the frequency, f_2, of a second tone so that it appears to be an octave higher in pitch, they will generally adjust f_2 to be roughly twice f_1.

However, when f_1 lies above 2.5 kHz, so that f_2 would lie above 5 kHz, octave matches become very erratic (Ward 1954). It appears that the musical interval of an octave is only clearly perceived when both tones are below 5 kHz.

Other aspects of the perception of pitch also change above 5 kHz. A sequence of pure tones above 5 kHz does not produce a clear sense of melody. This has been confirmed by experiments involving musical transposition. For example, Attneave and Olson (1971) asked subjects to reproduce sequences of tones (e.g., the NBC chimes) at different points along the frequency scale. Their results showed an abrupt breakpoint at about 5 kHz, above which transposition behavior was erratic. Also, subjects with absolute pitch (the ability to assign name notes without reference to other notes) are very poor at naming notes above 4 to 5 kHz (Ohgushi and Hatoh 1992).

These results are consistent with the idea that the pitch of pure tones is determined by different mechanisms above and below 5 kHz, specifically, by temporal mechanisms at low frequencies and place mechanisms at high frequencies. It appears that the perceptual dimension of tone height persists over the whole audible frequency range, but tone chroma only occurs in the frequency range below 5 kHz.

13.3 The Variation of Pitch with Level

The pitch of a pure tone is primarily determined by its frequency; however, sound level also plays a small role. On average, the pitch of tones below about 2000 Hz decreases with increasing level, while the pitch of tones above about 4000 Hz increases with increasing sound level. The early data of Stevens (1935) showed rather large effects of sound level on pitch, but more recent data generally show much smaller effects (e.g., Verschuure and van Meeteren 1975). For tones between 1 and 2 kHz, changes in pitch with level are generally less than 1%. For tones of lower and higher frequencies, the changes can be larger (up to 5%). There are also considerable individual differences both in the size of the pitch shifts with level and in the direction of the shifts.

It has sometimes been argued that pitch shifts with level are inconsistent with the temporal theory of pitch; neural interspike intervals are hardly affected by changes in sound level over a wide range. However, changes in pitch with level could be explained by the place theory, if shifts in level were accompanied by shifts in the position of maximum excitation on the basilar membrane. On closer examination, these arguments turn out to be rather weak. Although the temporal theory assumes that pitch depends on the temporal pattern of nerve spikes, it also assumes that the temporal information has to be "decoded" at some level in the auditory system. In other words, the time intervals between neural spikes have to be measured. It is quite possible that the mechanism which does this is affected by which neurons are active and by the spike rates in those neurons; these in turn depend on sound level.

The argument favoring the place mechanism is also weak. There is evidence from studies of auditory fatigue (temporary threshold shift) that the peak in the pattern of excitation evoked by low-frequency tones (around 1 kHz) shifts towards the base of the cochlea with increasing sound level (McFadden 1986). The base is tuned to higher frequencies, so the basal shift should correspond to hearing an increase in pitch. At high sound levels, the basal shift corresponds to a shift in frequency of one-half octave or more. Thus, the place theory predicts that the pitch of a 1-kHz tone should increase with increasing sound level, and the shift should correspond to half an octave or more at high sound levels. In fact, the pitch tends to decrease with increasing sound level, and the shift in pitch is always much less than half an octave.

13.4 General Conclusions on the Pitch Perception of Pure Tones

Several lines of evidence suggest that place mechanisms are not adequate to explain the frequency discrimination of pure tones. Contrary to the predictions of place theories, the DLF does not vary with frequency in the same way as the CB. Also, the DLF for short-duration tones is smaller than predicted by place theories except for frequencies above 5 kHz. This suggests that the DLF is determined by temporal mechanisms at low frequencies and by place mechanisms at high frequencies. The perception of sequences of pure tones also changes above 4 to 5 kHz. It seems that a sequence of tones only evokes a sense of musical interval or melody when the tones lie below 4 to 5 kHz, in the frequency range where temporal mechanisms probably operate.

14. The Pitch Perception of Complex Tones

14.1 The Phenomenon of the Missing Fundamental

The classical place theory has difficulty in accounting for the perception of complex tones. For such tones, the pitch does not, in general, correspond to the position of maximum excitation on the basilar membrane. A striking illustration of this is provided by the "phenomenon of the missing fundamental." Consider, as an example, a sound consisting of short impulses (clicks) occurring 200 times per second. This sound has a low pitch, which is very close to the pitch of a 200-Hz pure tone, and a sharp timbre. It contains harmonics with frequencies 200, 400, 600, 800, etc. Hz. However, it is possible to filter the sound so as to remove the 200-Hz component, and it is found that the pitch does not alter; the only result is a slight change in the timbre of the note. Indeed, all except a small group of mid-frequency harmonics can be eliminated, and the low pitch still remains, although the timbre becomes markedly different (see Yost and Shelft, Chapter 6).

Schouten (1970) called this low pitch associated with a group of high harmonics the "residue." He pointed out that the residue is distinguishable, subjectively, from a fundamental component which is physically presented or from a fundamental which may be generated (at high sound pressure levels) by nonlinear distortion in the ear. Thus, it seems that the perception of a residue pitch does not require activity at the point on the basilar membrane which would respond maximally to a pure tone of similar pitch. This is confirmed by a demonstration of Licklider (1956) that the low pitch of the residue persists even when low-frequency noise that would mask any component at the fundamental frequency is present. Several other names have been used to describe residue pitch, including "periodicity pitch," "virtual pitch" and "low pitch." This chapter will use the term residue pitch. Even when the fundamental component of a complex tone is present, the pitch of the tone is usually determined by harmonics other than the fundamental. Thus, the perception of a residue pitch should not be regarded as unusual. Rather, residue pitches are normally heard when listening to complex tones.

14.2 Discrimination of the Pitch of Complex Tones

When the repetition rate of a complex tone changes, all of the components change in frequency by the same ratio, and a change in residue pitch is heard. The ability to detect such changes in pitch is better than the ability to detect changes in a sinusoid at the fundamental frequency (Flanagan and Saslow 1958) and can be better than the ability to detect changes in the frequency of any of the sinusoidal components in the complex tone (Moore, Glasberg, and Shailer 1984). This indicates that information from the different harmonics is combined or integrated in the determination of residue pitch, which can lead to very fine discrimination; changes in repetition rate of about 0.2% can often be detected for fundamental frequencies in the range 100 to 400 Hz.

14.3 Analysis of a Complex Tone in the Peripheral Auditory System

A simulation of the analysis of a complex tone in the peripheral auditory system is illustrated in Figure 3.20. In this example, the complex tone is a periodic pulse train containing many equal-amplitude harmonics. The lower harmonics are partly resolved on the basilar membrane and give rise to distinct peaks in the pattern of activity along the basilar membrane. At a place tuned to the frequency of a low harmonic, the waveform on the basilar membrane is approximately a sine wave at the harmonic frequency. In contrast, the higher harmonics are not resolved and do not give rise to distinct peaks on the basilar membrane. The waveform at places on the membrane responding to higher harmonics is complex and has a repetition rate equal to the fundamental frequency of the sound.

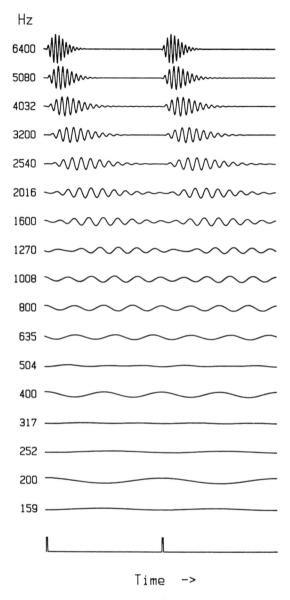

FIGURE 3.20. A simulation of responses on the basilar membrane to periodic impulses of 200 pulses per second. Each number on the left represents the frequency which would maximally excite a given point on the basilar membrane. The waveform which would be observed at that point, as a function of time, is plotted opposite that number. (From Moore 1989.)

15. Theories of Pitch Perception for Complex Tones

Several theories have been proposed to account for residue pitch. Theories prior to 1980 may be divided into two broad classes. The first, spectral theories, propose that the perception of the pitch of a complex tone involves two stages. The first stage is a frequency analysis which determines the frequencies of some of the individual sinusoidal components of the complex tone. The second stage is a pattern recognizer which determines the pitch of the complex from the frequencies of the resolved components (Goldstein 1973; Terhardt 1974). In essence, the pattern recognizer tries to find the harmonic series giving the best match to the resolved frequency components; the fundamental frequency of this harmonic series determines the perceived pitch. For these theories, the lower resolvable harmonics should determine the pitch that is heard.

The alternative, temporal theories, assume that pitch is based on the time pattern of the waveform at a point on the basilar membrane responding to the higher harmonics. Pitch is assumed to be related to the time interval between corresponding points in the fine structure of the waveform close to adjacent envelope maxima (Schouten, Ritsma, and Cardozo 1962). Nerve firings tend to occur at these points (i.e., phase locking occurs), so this time interval will be present in the time pattern of neural impulses. For these theories, the upper unresolved harmonics should determine the pitch that is heard.

Some recent theories (spectro-temporal theories) assume that both frequency analysis (spectral resolution) and time-pattern analysis are involved in pitch perception (Moore 1982, 1989; Srulovicz and Goldstein 1983; Patterson 1987; Yost and Sheft, Chapter 6).

16. Physical Variables Influencing Residue Pitch

16.1 Pitch of Inharmonic Complex Tones

Schouten, Ritsma, and Cardozo (1962) investigated the pitch of AM sine waves. If a carrier of frequency f_c is amplitude modulated by a modulator with frequency g, then the modulated wave contains components with frequencies $f_c - g$, f_c, and $f_c + g$. For example, a 2000-Hz carrier modulated 200 times per second contains components at 1800, 2000, and 2200 Hz and has a pitch similar to that of a 200-Hz sine wave.

Consider the effect of shifting the carrier frequency to, say, 2040 Hz. The complex now contains components at 1840, 2040, and 2240 Hz, which do not form a simple harmonic series. The perceived pitch, in this case, corresponds roughly to that of a 204-Hz sinusoid. The shift in pitch associated with a shift in carrier frequency is often called the "pitch shift of the residue." The shift demonstrates that the pitch of a complex tone is not determined by the envelope repetition rate, which is 200 Hz, nor by the spacing between adjacent components, which is also 200 Hz. In addition, there is an ambiguity of pitch; pitches around 185 Hz and 227 Hz may also be heard.

The perceived pitch can be explained by both spectral theories and temporal theories. According to the spectral theories, a harmonic complex tone with a fundamental of 204 Hz would provide a good "match"; this would have components at 1836, 2040, and 2244 Hz. Errors in estimating the fundamental occur mainly through errors in estimating the appropriate harmonic number. In the above example, the presented components were assumed by the pattern recognizer to be the 9th, 10th, and 11th harmonics of a 204-Hz fundamental. However, a reasonable fit could also be found by assuming the components to be the 8th, 9th, and 10th harmonics of a 226.7-Hz fundamental or the 10th, 11th, and 12th harmonics of a 185.5-Hz fundamental. Thus, the theories predict the ambiguities of pitch which are actually observed for this stimulus.

The pitch shift of the residue is explained by temporal theories in the

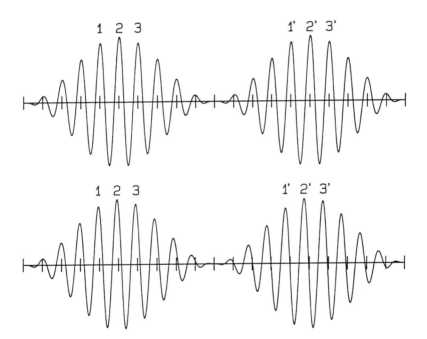

FIGURE 3.21. Waveforms for two amplitude-modulated sine waves. In the upper waveform, the carrier frequency (2000 Hz) is an exact multiple of the modulation frequency (200 Hz). Thus, the time intervals between corresponding peaks in the fine structure, 1-1', 2-2', 3-3', are 5 ms, the period of the modulation frequency. In the lower waveform, the carrier frequency (2040 Hz) is shifted slightly upwards. Thus, 10 complete cycles of the carrier occur in slightly less than 5 ms, and the time intervals between corresponding points on the waveform, 1-1', 2-2', 3-3', are slightly shorter than 5 ms. The lower waveform has a slightly higher pitch than the upper.

following way. For the complex tone with a shifted carrier frequency, the time intervals between corresponding peaks in the fine structure of the wave are slightly less than 5 ms; roughly ten periods of the carrier occur for each period of the envelope, but the exact time for ten periods of a 2040-Hz carrier is 4.9 ms, which corresponds to a pitch of 204 Hz. This is illustrated in Figure 3.21. Nerve spikes can occur at any of the prominent peaks in the waveform, labeled 1, 2, 3 and 1', 2', 3'. Thus, the time intervals between successive nerve spikes fall into several groups corresponding to the intervals 1-1', 1-2', 1-3', 2-1', etc. This can account for the fact that the pitches of these stimuli are ambiguous. The intervals 1-1', 2-2', and 3-3' correspond to the most prominent pitch of about 204 Hz, while other intervals, such as 1-2' and 2-1', correspond to the other pitches which may be perceived.

16.2 The "Existence Region" of the Tonal Residue

According to the spectral theories, a residue pitch should only be heard when at least one of the sinusoidal components of the stimulus can be resolved. Thus, a residue pitch should not be heard for stimuli containing only very high unresolvable harmonics. Ritsma (1962) investigated the audibility of residue pitches as a function of the modulation rate and the harmonic number (f_c/g) for AM sinusoids containing three components. Later, he extended the results to complexes with a greater number of components (Ritsma 1963). He found that the tonal character of the residue pitch existed only within a limited frequency region, referred to as the "existence region." When the harmonic number was too high (above about 20), only a kind of high buzz of undefinable pitch was heard. However, residue pitches could clearly be heard for harmonic numbers between 10 and 20. These harmonics would not be resolvable.

Ritsma's results may have been influenced by the presence of combination tones in the frequency region below the lowest components in the complex tones, as has since been pointed out by Ritsma himself (1970). However, more recent experiments that have made use of background noise to mask combination tones also indicate that residue pitch can be perceived when no resolvable harmonics are present (Moore 1973b). Moore and Rosen (1979) and Houtsma and Smurzynski (1990) have demonstrated that stimuli containing only high, unresolvable harmonics can evoke a sense of musical interval and, hence, of pitch. The spectral theories cannot account for this.

16.3 The Principle of Dominance

Ritsma (1967) carried out an experiment to determine which components in a complex sound are most important in determining its pitch. He presented complex tones in which the frequencies of a small group of harmonics were multiples of a fundamental that was slightly higher or lower than the fundamental of the remainder. The subject's pitch judgments were used to determine whether the pitch of the complex as a whole was affected by the shift in the group of harmonics. Ritsma found that "for fundamental frequencies in the range 100 Hz to 400 Hz and for sensation levels up to at least 50 dB above threshold of the entire signal, the frequency band consisting of the third, fourth, and fifth harmonics tends to dominate the pitch sensation as long as its amplitude exceeds a minimum absolute level of about 10 dB above threshold."

Thus, Ritsma (1970) introduced the concept of dominance: "If pitch information is available along a large part of the basilar membrane, the ear uses only the information from a narrow band. This band is positioned at 3 to 5 times the pitch value. Its precise position depends somewhat on the subject."

This finding has been broadly confirmed in other ways (Plomp 1967), although the data of Moore, Glasberg, and Shailer (1984) and Moore, Glasberg, and Peters (1985) show that there are large individual differences in which harmonics are dominant, and, for some subjects, the first two harmonics play an important role. Other data also show that the dominant region is not fixed in terms of harmonic number but depends somewhat on absolute frequency (Plomp 1967; Patterson and Wightman 1976). For high fundamental frequencies (above about 1000 Hz), the fundamental is usually the dominant component, while, for very low fundamental frequencies, around 50 Hz, harmonics above the fifth may be dominant (Moore and Glasberg 1988; Moore and Peters 1992).

On the whole, these results are better explained by the spectral theories. The 3rd, 4th, and 5th harmonics are relatively well resolved on the basilar membrane and are usually separately perceptible. However, for very low fundamental frequencies, around 50 Hz, harmonics in the dominant region (above the 5th harmonic) would not be resolvable.

16.4 *Pitch of Dichotic Two-Tone Complexes*

Houtsma and Goldstein (1972) investigated the perception of stimuli which comprised just two harmonics with successive harmonic numbers, e.g., the fourth and fifth harmonics of a (missing) fundamental. They found that musically trained subjects were able to identify musical intervals between such stimuli when they were presented dichotically (one harmonic to each ear), presumably indicating that residue pitches were heard. These results cannot be explained by temporal theories, which require an interaction of components on the basilar membrane. They imply that "the pitch of these complex tones is mediated by a central processor operating on neural signals derived from those effective stimulus harmonics which are tonotopically resolved" (Houtsma and Goldstein 1972). However, it is not yet entirely clear that the findings apply to the pitches of complex tones in general. The pitch of a two-tone complex is not very clear; indeed, many observers do not hear a single low pitch but rather perceive the component tones individually (Smoorenburg 1970).

16.5 Pitch of Rippled Noise

A residue-type pitch can be heard not only for stimuli containing discrete sinusoidal components, but also for noise-like stimuli with regularly spaced spectral peaks. For example, the rippled-noise stimulus described in section 3.3 evokes a pitch related to the delay time, T (Fourcin 1965; Bilsen 1966; Yost, Hill, and Perez-Falcon 1978). This pitch is sometimes described as "repetition pitch." For the cosine+ stimulus, the pitch corresponds to $1/T$ Hz. However, for the cosine- stimulus, the pitch is somewhat weaker, is ambiguous, and is judged to be either slightly higher or lower than that corresponding to $1/T$; pitch matches are usually around $1.14/T$ and $0.89/T$.

The cosine- stimulus may be considered as analogous to some of the inharmonic stimuli described in Section 16.1, except that the spectrum is continuous rather than discrete. The theories used to explain the pitch of inharmonic stimuli in Section 16.1 can also be applied to the cosine- stimulus. However, to predict the correct pitch values, it is necessary to assume that there is a dominant region; the pitch appears to be determined by information in a spectral region centered at three to five times the frequency value corresponding to the perceived pitch (Bilsen 1966; Yost and Hill 1979). This corresponds rather well to the dominant region found for complex tones with discrete sinusoidal components (Section 16.3). Yost and Hill (1979) and Yost (1982) have suggested that the dominant region can be described by a spectral weighting function and that this function depends partly on lateral suppression.

16.6 Pitch of Stimuli Without Spectral Cues

A number of workers have investigated pitches evoked by stimuli that contain no spectral peaks and therefore would not produce a well-defined maximum on the basilar membrane. It has been argued that such pitches cannot arise on the basis of place information and so provide evidence for the operation of temporal mechanisms.

Miller and Taylor (1948) used random white noise that was turned on and off abruptly and periodically, at various rates. Such interrupted noise has a flat long-term magnitude spectrum. They reported that the noise had a pitchlike quality for interruption rates between 100 and 250 Hz. Burns and Viemeister (1976, 1981) showed that noise that was sinusoidally AM had a pitch corresponding to the modulation rate. As for interrupted noise, the long-term magnitude spectrum of white noise remains flat after modulation. They found that the pitch could be heard for modulation rates up to 800 to 1000 Hz and that subjects could recognize musical intervals and melodies played by varying the modulation rate.

It should be noted that, although the long-term magnitude spectrum of interrupted or AM noise is flat, some spectral cues exist in the short-term magnitude spectrum (Pierce, Lipes, and Cheetham 1977). Thus, the use of place cues to extract the pitch of these stimuli cannot be completely ruled out.

This is not the case, however, for experiments showing pitch sensations as a result of the binaural interaction of noise stimuli. Huggins and Cramer (1958) fed white noise from the same noise generator to both ears via headphones. The noise to one ear went through an all-pass filter which passed all audible frequency components without change of amplitude, but produced a phase change in a small frequency region. A faint pitch was heard corresponding to the frequency region of the phase transition, although the stimulus presented to each ear alone was white noise and produced no pitch sensation. The effect was only heard when the phase change was in the frequency range below about 1 kHz. A phase shift of a narrow band of the noise results in that band being heard in a different position in space from the rest of the noise. This spatial separation presumably produces the pitchlike character associated with that narrow band. Pitches produced by binaural interaction have also been reported by Fourcin (1970), Bilsen and Goldstein (1974), Kubovy, Cutting, and McGuire (1974), Nein and Hartmann (1981), Yost, Harder, and Dye (1987), and Yost and Sheft (Chapter 6).

16.7 Effect of the Relative Phases of the Components on Pitch

Changes in the relative phases of the components making up a complex tone change the temporal structure of the tone, but have little effect on the auditory representation of the lower, resolved harmonics. Thus, spectral theories predict that relative phase should not influence pitch, while temporal theories predict that relative phase should affect pitch; pitch should be more distinct for phase relations producing "peaky" waveforms after auditory filtering (waveforms with high crest factors). For tones containing many harmonics, phase has only small effects on the value of the pitch heard (Patterson 1973), but it does affect the clarity of pitch (Lundeen and Small 1984). For tones containing only a few high harmonics, phase can affect both the pitch value and clarity of the pitch (Moore 1977). Phase can affect the discrimination of changes in residue pitch, and the phase effects are larger in hearing-impaired subjects than in normal subjects, presumably because auditory filters are broader in impaired subjects and produce more interaction of the components in the peripheral auditory system (Moore and Peters 1992). Overall, these results suggest that the temporal structure of complex sounds does influence pitch perception.

17. Spectro-Temporal Models of the Pitch Perception of Complex Tones

The evidence reviewed in Section 16 indicates that neither spectral nor temporal theories can account for all aspects of the pitch perception of complex tones. Spectral theories cannot account for the following facts: a residue pitch can be perceived when only high, unresolvable harmonics are present; stimuli without spectral peaks can give rise to a pitch sensation; and the relative phases of the components can affect pitch perception. Temporal theories cannot account for these facts: the lower, resolvable harmonics are dominant in determining pitch; and a residue pitch may be heard when there is no possibility of the components interacting in the peripheral auditory system.

These findings have led several workers to propose theories in which both spectral and temporal mechanisms play a role; the initial place/spectral analysis in the cochlea is followed by an analysis of the time pattern of the waveform at each place or of the neural spikes evoked at each place (Moore 1982, 1989; Moore and Glasberg 1986; Patterson 1987; Meddis and Hewitt 1991a,b). The model proposed by Moore (1982, 1989) is illustrated in Figure 3.22. The first stage in the model is a bank of bandpass filters with continu-

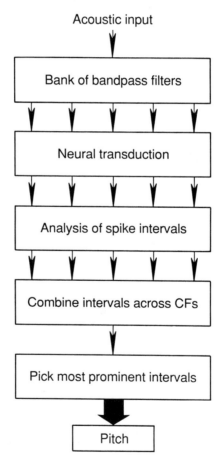

FIGURE 3.22. A spectro-temporal model for the perception of the pitch of complex tones. (From Moore 1989.)

ously overlapping passbands. These are the auditory filters or critical bands. The outputs of the filters in response to a complex tone have the form shown in Figure 3.20. The filters responding to low harmonics have outputs which are approximately sinusoidal in form; the individual harmonics are resolved. The filters responding to higher harmonics have outputs corresponding to the interaction of several harmonics. The waveform is complex but has a repetition rate corresponding to that of the input.

The next stage in the model is the transduction of the filter outputs to neural impulses. The temporal pattern of firing in a given neuron reflects the temporal structure of the waveform driving that neuron. Say, for example, that the input has a fundamental frequency of 200 Hz. The fourth harmonic, at 800 Hz, is well resolved in the filter bank, and hence the neurons with CFs close to 800 Hz respond as if the input were an 800-Hz sinusoid. The time intervals between successive nerve impulses are multiples of the period of that tone, i.e., 1.25, 2.5, 3.75, 5.0, ... ms. Neurons with higher CFs, say, around 2000 Hz, are driven by a more complex waveform. The temporal structure of the response is correspondingly complex. Each peak in the fine structure of the waveform is capable of evoking a spike, so that many different time intervals occur between successive spikes. The interval corresponding to the fundamental, 5 ms, is present, but other intervals, such as 4.0, 4.5, 5.5, and 6.0 ms, also occur (Evans 1978; Javel 1980).

The next stage in the model is a device which analyzes, separately for each CF, the interspike intervals that are present. The range of intervals which can be analyzed is probably limited and varies with CF. At a given CF, the device probably operates over a range from about 0.5/CF to 15/CF seconds. This range is appropriate for the time intervals which would occur most often.

The next stage is a device that compares the time intervals present in the different channels and searches for common time intervals. The device may also integrate information over time. In general, the time interval that is found most often corresponds to the period of the fundamental component. Finally, the time intervals that are most prominently represented across channels are fed to a decision mechanism which selects one interval from among those passed to it. The perceived pitch corresponds to the reciprocal of the final interval selected.

Consider how this model deals with a complex tone composed of a few low harmonics, say, the 3rd, 4th, and 5th harmonics of a 200-Hz (missing) fundamental. In those neurons responding primarily to the 600-Hz component, an analysis of time intervals between successive nerve firings would reveal intervals of 1.67, 3.33, 5.0, 6.67, ... ms, each of which is an integral multiple of the period of that component. Similarly, the intervals in the 800-Hz "channel" are 1.25, 2.5, 3.75, 5.0, ... ms, while those in the 1000-Hz "channel" are 1, 2, 3, 4, 5, ... ms. The only time interval which is in common across all of the channels is 5 ms; this corresponds to the pitch of the missing fundamental. Thus, this stimulus will evoke a clear, unambiguous pitch.

Consider next what happens when a small group of high harmonics is presented, say, the 12th, 13th, and 14th harmonics of 200 Hz. These harmonics are not resolved in the filter bank, and so essentially only one channel of timing information is available. Furthermore, the time interval information is ambiguous with many values closely spaced around 5 ms being present. Thus, while a pitch corresponding to the missing fundamental may be perceived, that pitch is weak and ambiguous, as is observed psychophysically. This can explain why lower harmonics tend to dominate in the perception of pitch; the temporal information they provide is far less ambiguous, provided that information is combined across channels. The pitch associated with high harmonics can be made clear by increasing the number of harmonics, since the ambiguities can be reduced by comparing across channels. This also is observed psychophysically.

If a small group of very high harmonics is presented, then they may fail to evoke a sense of musical pitch. This can be explained in terms of the limited range of time intervals that can be analyzed at each CF. For harmonics above about the 15th, the time interval corresponding to the fundamental falls outside the range that can be analyzed in the channel responding to those harmonics. This is one factor contributing to the limited existence region of the tonal residue. The other factor is the absolute upper limit for phase locking. When the harmonics lie above 5 kHz, the fine structure of the waveform at the filter output is no longer preserved in the temporal patterning of neural impulses. Thus, the later stages of the analysis of interspike intervals do not reveal the regularities necessary to determine the fundamental.

The way that the model deals with the pitch of nonharmonic complex tones depends on the spacing of the components relative to their center frequency. If the components are widely spaced, and therefore resolvable, the pitch of the complex is determined in a similar way to that proposed by Terhardt (1974). Say, for example, that components at 840, 1040, and 1240 Hz are presented. The time intervals in the 840-Hz channel will be 1.19, 2.38, 3.57, 4.76, 5.95, ... ms, all of which are integral multiples of the period 1.19 ms. Similarly, the intervals in the 1040-Hz channel are 0.96, 1.92, 2.88, 3.85, 4.8, 5.76, ... ms, and those in the 1240-Hz channel are 0.81, 1.61, 2.42, 3.22, 4.03, 4.84, ... ms. In this case, there is no time interval that is exactly the same in all channels, but there is a near coincidence at 4.8 ms, the intervals in the three channels being 4.76, 4.8, and 4.84 ms. The perceived pitch thus corresponds to the reciprocal of 4.8 ms, i.e., 208 Hz. When the components are closely spaced, and therefore unresolvable, then the pitch is derived from the time interval that is most prominently represented in the pattern of neural impulses evoked by the complex. In this case, the model works in the same way as the temporal model described in Section 16.1. The perceived pitch corresponds to the time interval between peaks in the fine structure of the waveform (at the output of the auditory filter) close to adjacent envelope maxima. The pitch is ambiguous since several "candidate" time intervals may be present.

In this model, just as in the other models discussed, combination tones may play a role in determining the pitch percept. Their role may be particularly important when the stimulus itself contains only closely spaced components. The combination tones act like lower partials and are more resolvable than the partials physically present in the stimulus.

The pitches of stimuli without spectral peaks, such as periodically interrupted noise, are explained by the model as arising from the time interval information present primarily in the channels with high CFs, where the filter bandwidths are wider. When a filter has a wide bandwidth, the temporal structure of the input tends to be preserved at the output. Thus, the temporal patterns of firing reflect the interruption of the noise. However, the time intervals between successive nerve impulses are much less regular than for a periodic sound, since the exact waveform of the noise varies randomly from moment to moment; the only regularity is in the timing of the envelope, not the fine structure. Thus, the pitch of interrupted noise is weak.

The model as presented does not account for the pitches which arise from a combination of the information at the two ears. However, it can easily be extended to do this by making the following reasonable assumptions. There is a separate filter bank, transduction mechanism, and interval analyzer for each ear. There is also a device determining interaural time intervals for spikes in channels with corresponding CFs. The results of the separate interval analysis for each ear, and of the interaural interval analysis, are fed to a common central interval comparator and decision mechanism.

In summary, the model can account for the major features of the pitch perception of complex tones, including the dominant region, the existence region of the residue, the pitch of nonharmonic complexes, and pitches produced by binaural interaction. Furthermore, it is consistent with the evidence presented in Sections 13.1 to 13.3 that the pitch of pure tones is determined primarily by temporal mechanisms for frequencies up to 5 kHz. Notice, however, that both place and temporal analysis play a crucial role in the model; neither alone is sufficient.

18. Summary

The peripheral auditory system contains a bank of band-pass filters, the auditory filters, with continuously overlapping passbands. The basilar membrane appears to provide the initial basis of the filtering process. The auditory filter can be thought of as a weighting function which characterizes frequency selectivity at a particular center frequency. Its bandwidth for frequencies above 1 kHz is about 10% to 17% of the center frequency. At moderate sound levels, the auditory filter is roughly symmetrical on a linear frequency scale. At high sound levels, the low-frequency side of the filter becomes less steep than the high-frequency side. The critical bandwidth is related to the bandwidth of the auditory filter. It is revealed in experiments on masking, loudness, absolute threshold, phase sensitivity, and the audibility of partials in complex tones.

The excitation pattern of a given sound represents the distribution of activity evoked by that sound as a function of the CF of the neurons stimulated. In psychophysical terms, the excitation pattern can be defined as the output of each auditory filter as a function of its center frequency. The shapes of excitation patterns for sinusoids or narrowband noises are similar to the masking patterns of narrowband noises.

Houtgast (1974) and others have shown that nonsimultaneous masking reveals suppression effects similar to the suppression observed in primary auditory neurons. This suppression is not revealed in simultaneous masking, possibly because suppression at a given CF does not affect the signal-to-masker ratio at that CF. One result of suppression is an enhancement in frequency selectivity.

In principle, there are two ways in which the pitch of a sound may be coded; by the distribution of activity across different auditory neurons and by the temporal patterns of firing within and across neurons. It is likely that both types of information are utilized but that their relative importance is different for different frequency ranges and for different types of sounds.

The neurophysiological evidence in animals indicates that the synchrony of nerve impulses to a particular phase of the stimulating waveform disappears above 4 to 5 kHz. Above this frequency, the ability to discriminate changes in the frequency of pure tones diminishes, and the sense of musical pitch disappears. It is likely that this reflects the use of temporal information in the frequency range below 4 to 5 kHz.

For complex tones, two classes of pitch theories have been popular. Temporal theories suggest that the pitch of a complex tone is derived from the time intervals between successive nerve firings evoked at a point on the basilar membrane where adjacent partials are interfering with one another. Spectral theories suggest that pitch is derived by a central processor operating on neural signals corresponding to the individual partials present in the complex sound. In both theories, combination tones in the frequency region below the lowest partial may influence the pitch percept.

Spectral theories are supported by these findings: low, resolvable harmonics tend to dominate in the perception of pitch; and a pitch corresponding to a "missing fundamental" can be perceived when there is no possibility of an interference of partials on the basilar membrane. Temporal theories are supported by these findings: (weak) pitches can be perceived when the harmonics are too close in frequency to be resolvable and when the stimuli have no well-defined spectral structure (e.g., interrupted noise); and pitch perception and discrimination can be affected by the relative phases of the components in a complex tone. Spectro-temporal models incorporate features of both the temporal and the spectral models. These models can account for most experimental data on the pitch perception of complex tones.

Acknowledgements. I thank Brian Glasberg and Michael Shailer for helpful comments on an earlier version of this chapter.

References

American National Standards Institute (1973) American national psychoacoustical terminology. S3.20. New York.

Arthur RM, Pfeiffer RR, Suga N (1971) Properties of "two-tone inhibition" in primary auditory neurones. J Physiol 212:593-609.

Attneave F, Olson RK (1971) Pitch as a medium: A new approach to psychophysical scaling. Am J Psychol 84:147-166.

Bachem A (1950) Tone height and tone chroma as two different pitch qualities. Acta Psychol 7:80-88.

Bacon SP, Moore BCJ (1986) Temporal effects in masking and their influence on psychophysical tuning curves. J Acoust Soc Am 80:1638-1654.

Bacon SP, Viemeister NF (1985) The temporal course of simultaneous tone-on-tone masking. J Acoust Soc Am 78:1231-1235.

Bilsen FA (1966) Repetition pitch: Monaural interaction of a sound with the same, but phase shifted, sound. Acustica 17:265-300.

Bilsen FA, Goldstein JL (1974) Pitch of dichotically delayed noise and its possible spectral basis. J Acoust Soc Am 55:292-296.

Burns EM, Viemeister NF (1976) Nonspectral pitch. J Acoust Soc Am 60:863-869.

Burns EM, Viemeister NF (1981) Played again SAM: Further observations on the pitch of amplitude-modulated noise. J Acoust Soc Am 70:1655-1660.

Buus S (1985) Release from masking caused by envelope fluctuations. J Acoust Soc Am 78:1958-1965.

Buus S, Schorer E, Florentine M, Zwicker E (1986) Decision rules in detection of simple and complex tones. J Acoust Soc Am 80:1646-1657.

Chistovich LA (1957) Frequency characteristics of masking effect. Biophys 2:743-755.

Coninx F (1978) The detection of combined differences in frequency and intensity. Acustica 39:137-150.

Delgutte B (1988) Physiological mechanisms of masking. In: Duifhuis H, Horst JW, Wit HP (eds) Basic Issues in Hearing. London: Academic Press, pp. 204-212.

Delgutte B (1990) Physiological mechanisms of psychophysical masking: Observations from auditory-nerve fibers. J Acoust Soc Am 87:791-809.

Egan JP, Hake HW (1950) On the masking pattern of a simple auditory stimulus. J Acoust Soc Am 22:622-630.

Emmerich DS, Ellermeier W, Butensky B (1989) A re-examination of the frequency discrimination of random-amplitude tones, and a test of Henning's modified energy-detector model. J Acoust Soc Am 85:1653-1659.

Evans EF (1978) Place and time coding of frequency in the peripheral auditory system: Some physiological pros and cons. Audiology 17:369-420.

Evans EF, Pratt SR, Cooper NP (1989) Correspondence between behavioural and physiological frequency selectivity in the guinea pig. Br J Audiol 23:151-152.

Feth LL (1972) Combinations of amplitude and frequency differences in auditory discrimination. Acustica 26:67-77.

Flanagan JL, Saslow MG (1958) Pitch discrimination for synthetic vowels. J Acoust Soc Am 30:435-442.

Fletcher H (1940) Auditory patterns. Rev Mod Phys 12:47-65.

Florentine M, Buus S (1981) An excitation-pattern model for intensity discrimination. J Acoust Soc Am 70:1646-1654.

Fourcin AJ (1965) The pitch of noise with periodic spectral peaks. Fifth Int Cong Acoust Ia, B 42.

Fourcin AJ (1970) Central pitch and auditory lateralization. In: Plomp R, Smoorenburg GF (eds) Frequency Analysis and Periodicity Detection in Hearing. Leiden, The Netherlands: AW Sijthoff, pp. 319–328.

Gässler G (1954) Über die Horschwelle fur Schallereignisse mit verschieden breitem Frequenzspektrum. Acustica 4:408–414.

Glasberg BR, Moore BCJ (1986) Auditory filter shapes in subjects with unilateral and bilateral cochlear impairments. J Acoust Soc Am 79:1020–1033.

Glasberg BR, Moore BCJ (1990) Derivation of auditory filter shapes from notched-noise data. Hear Res 47:103–138.

Glasberg BR, Moore BCJ, Nimmo-Smith I (1984) Comparison of auditory filter shapes derived with three different maskers. J Acoust Soc Am 75:536–546.

Glasberg BR, Moore BCJ, Patterson RD, Nimmo-Smith I (1984) Dynamic range and asymmetry of the auditory filter. J Acoust Soc Am 76:419–427.

Goldstein JL (1973) An optimum processor theory for the central formation of the pitch of complex tones. J Acoust Soc Am 54:1496–1516.

Goldstein JL, Srulovicz P (1977) Auditory-nerve spike intervals as an adequate basis for aural frequency measurement. In: Evans EF, Wilson JP (eds) Psychophysics and Physiology of Hearing. London: Academic Press, pp. 337–346.

Greenwood DD (1961) Critical bandwidth and the frequency coordinates of the basilar membrane. J Acoust Soc Am 33:1344–1356.

Greenwood DD (1971) Aural combination tones and auditory masking. J Acoust Soc Am 50:502–543.

Greenwood DD (1990) A cochlear frequency-position function for several species—29 years later. J Acoust Soc Am 87:2592–2605.

Hall JW III, Haggard MP, Fernandes MA (1984) Detection in noise by spectro-temporal pattern analysis. J Acoust Soc Am 76:50–56.

Hamilton PM (1957) Noise masked threshold as a function of tonal duration and masking noise bandwidth. J Acoust Soc Am 29:506–511.

Henning GB (1966) Frequency discrimination of random amplitude tones. J Acoust Soc Am 39:336–339.

Henning GB (1967) A model for auditory discrimination and detection. J Acoust Soc Am 42:1325–1334.

Higgins MB, Turner CW (1990) Summation bandwidths at threshold in normal and hearing-impaired listeners. J Acoust Soc Am 88:2625–2630.

Houtgast T (1972) Psychophysical evidence for lateral inhibition in hearing. J Acoust Soc Am 51:1885–1894.

Houtgast T (1973) Psychophysical experiments on "tuning curves" and "two-tone inhibition." Acustica 29:168–179.

Houtgast T (1974) Lateral suppression in hearing. Ph.D. Thesis, Free University of Amsterdam. Amsterdam: Academische Pers. BV.

Houtgast T (1977) Auditory-filter characteristics derived from direct-masking data and pulsation-threshold data using a rippled-noise masker. J Acoust Soc Am 62:409–415.

Houtsma AJM, Goldstein JL (1972) The central origin of the pitch of complex tones: Evidence from musical interval recognition. J Acoust Soc Am 51:520–529.

Houtsma AJM, Smurzynski, J (1990) Pitch identification and discrimination for complex tones with many harmonics. J Acoust Soc Am 87:304–310.

Huggins WH, Cramer EM (1958) Creation of pitch through binaural interaction. J Acoust Soc Am 30:413–417.
Javel E (1980) Coding of AM tones in the chinchilla auditory nerve: Implications for the pitch of complex tones. J Acoust Soc Am 68:133–146.
Johnson-Davies DB, Patterson RD (1979) Psychophysical tuning curves: Restricting the listening band to the signal region. J Acoust Soc Am 65:765–770.
Klein MA, Hartmann WM (1981) Binaural edge pitch. J Acoust Soc Am 70:51–61.
Kubovy M, Cutting JE, McGuire RM (1974) Hearing with the third ear: Dichotic perception of a melody without monaural familiarity cues. Science 186:272–274.
Licklider JCR (1956) Auditory frequency analysis. In: Cherry C (ed) Information Theory. New York: Academic Press, pp. 253–268.
Lundeen C, Small AM (1984) The influence of temporal cues on the strength of periodicity pitches. J Acoust Soc Am 75:1578–1587.
Lutfi RA, Patterson RD (1984) On the growth of masking asymmetry with stimulus intensity. J Acoust Soc Am 76:739–745.
Maiwald D (1967) Die Berechnung von Modulationsschwellen mit Hilfe eines Funktionsschemas. Acustica 18:193–207.
McFadden DM (1986) The curious half-octave shift: Evidence for a basalward migration of the traveling-wave envelope with increasing intensity. In: Salvi RJ, Henderson D, Hamernik RP, Colletti V (eds) Basic and Applied Aspects of Noise-Induced Hearing Loss. New York: Plenum Press.
Meddis R, Hewitt M (1991a) Virtual pitch and phase sensitivity of a computer model of the auditory periphery. I: Pitch identification. J Acoust Soc Am 89:2866–2882.
Meddis R, Hewitt M (1991b) Virtual pitch and phase sensitivity of a computer model of the auditory periphery. II: Phase sensitivity. J Acoust Soc Am 89:2883–2894.
Miller GA, Taylor W (1948) The perception of repeated bursts of noise. J Acoust Soc Am 20:171–182.
Moore BCJ (1972). Some experiments relating to the perception of pure tones: Possible clinical applications. Sound 6:73–79.
Moore BCJ (1973a) Frequency difference limens for short-duration tones. J Acoust Soc Am 54:610–619.
Moore BCJ (1973b) Some experiments relating to the perception of complex tones. Quart J Exp Psychol 25:451–475.
Moore BCJ (1977) Effects of relative phase of the components on the pitch of three-component complex tones. In: Evans EF, Wilson JP (eds) Psychophysics and Physiology of Hearing. London: Academic Press, pp. 349–358.
Moore BCJ (1978) Psychophysical tuning curves measured in simultaneous and forward masking. J Acoust Soc Am 63:524–532.
Moore BCJ (1980) Detection cues in forward masking. In: van den Brink G, Bilsen FA (eds) Psychophysical, Physiological and Behavioural Studies in Hearing. Delft, The Netherlands: Delft University Press, pp. 222–229.
Moore BCJ (1981) Interactions of masker bandwidth with signal duration and delay in forward masking. J Acoust Soc Am 70:62–68.
Moore BCJ (1982) An Introduction to the Psychology of Hearing, Second Edition. London: Academic Press.
Moore BCJ (1986) Parallels between frequency selectivity measured psychophysically and in cochlear mechanics. Scand Audiol Suppl 25:139–152.
Moore BCJ (1987) Distribution of auditory-filter bandwidths at 2 kHz in young normal listeners. J Acoust Soc Am 81:1633–1635.

Moore BCJ (1989) An Introduction to the Psychology of Hearing, Third Edition. London: Academic Press.

Moore BCJ, Glasberg BR (1981) Auditory filter shapes derived in simultaneous and forward masking. J Acoust Soc Am 70:1003–1014.

Moore BCJ, Glasberg BR (1982a) Contralateral and ipsilateral cueing in forward masking. J Acoust Soc Am 71:942–945.

Moore BCJ, Glasberg BR (1982b) Interpreting the role of suppression in psychophysical tuning curves. J Acoust Soc Am 72:1374–1379.

Moore BCJ, Glasberg BR (1983) Suggested formulae for calculating auditory-filter bandwidths and excitation patterns. J Acoust Soc Am 74:750–753.

Moore BCJ, Glasberg BR (1985) The danger of using narrowband noise maskers to measure suppression. J Acoust Soc Am 77:2137–2141.

Moore BCJ, Glasberg BR (1986) The role of frequency selectivity in the perception of loudness, pitch and time. In: Moore BCJ (ed) Frequency Selectivity in Hearing. London: Academic Press, pp. 251–308.

Moore BCJ, Glasberg BR (1987) Formulae describing frequency selectivity as a function of frequency and level, and their use in calculating excitation patterns. Hear Res 28:209–225.

Moore BCJ, Glasberg BR (1988) Effects of the relative phase of the components on the pitch discrimination of complex tones by subjects with unilateral cochlear impairments. In: Duifhuis H, Horst JW, Wit HP (eds) Basic Issues in Hearing. London: Academic Press, pp. 421–430.

Moore BCJ, Glasberg BR (1989) Mechanisms underlying the frequency discrimination of pulsed tones and the detection of frequency modulation. J Acoust Soc Am 86:1722–1732.

Moore BCJ, O'Loughlin BJ (1986) The use of nonsimultaneous masking to measure frequency selectivity and suppression. In: Moore BCJ (ed) Frequency Selectivity in Hearing. London: Academic Press, pp. 179–250.

Moore BCJ, Peters RW (1992) Pitch discrimination and phase sensitivity in young and elderly subjects and its relationship to frequency selectivity. J Acoust Soc Am 91:2881–2893.

Moore BCJ, Rosen SM (1979) Tune recognition with reduced pitch and interval information. Quart J Exp Psychol 31:229–240.

Moore BCJ, Glasberg BR, Roberts B (1984) Refining the measurement of psychophysical tuning curves. J Acoust Soc Am 76:1057–1066.

Moore BCJ, Glasberg BR, Shailer MJ (1984) Frequency and intensity difference limens for harmonics within complex tones. J Acoust Soc Am 75:550–561.

Moore BCJ, Glasberg BR, Peters RW (1985) Relative dominance of individual partials in determining the pitch of complex tones. J Acoust Soc Am 77:1853–1860.

Moore BCJ, Poon PWF, Bacon SP, Glasberg BR (1987) The temporal course of masking and the auditory filter shape. J Acoust Soc Am 81:1873–1880.

Moore BCJ, Peters RW, Glasberg BR (1990) Auditory filter shapes at low center frequencies. J Acoust Soc Am 88:132–140.

Neff DL (1985) Stimulus parameters governing confusion effects in forward masking. J Acoust Soc Am 78:1966–1976.

Ohgushi K, Hatoh T (1992) The musical pitch of high frequency tones. In: Cazals Y, Demany L, Horner K (eds) Auditory Physiology and Perception. Oxford: Pergamon Press, pp. 207–212.

Ohm GS (1843) Uber die Definition des Tones, nebst daran geknüpfter Theorie der Sirene und ähnlicher tonbildender Vorrichtungen. Ann Phys Chem 59:513–565.

O'Loughlin BJ, Moore BCJ (1981) Improving psychoacoustical tuning curves. Hear Res 5:343–346.
Patterson RD (1973) The effects of relative phase and the number of components on residue pitch. J Acoust Soc Am 53:1565–1572.
Patterson RD (1974) Auditory filter shape. J Acoust Soc Am 55:802–809.
Patterson RD (1976) Auditory filter shapes derived with noise stimuli. J Acoust Soc Am 59:640–654.
Patterson RD (1987) A pulse ribbon model of monaural phase perception. J Acoust Soc Am 82:1560–1586.
Patterson RD, Henning GB (1977) Stimulus variability and auditory filter shape. J Acoust Soc Am 62:649–664.
Patterson RD, Moore BCJ (1986) Auditory filters and excitation patterns as representations of frequency resolution. In: Moore BCJ (ed) Frequency Selectivity in Hearing. London: Academic Press, pp. 123–177.
Patterson RD, Nimmo-Smith I (1980) Off-frequency listening and auditory-filter asymmetry. J Acoust Soc Am 67:229–245.
Patterson RD, Wightman FL (1976) Residue pitch as a function of component spacing. J Acoust Soc Am 59:1450–1459.
Patterson RD, Nimmo-Smith I, Weber DL, Milroy R (1982) The deterioration of hearing with age: Frequency selectivity, the critical ratio, the audiogram and speech threshold. J Acoust Soc Am 72:1788–1803.
Pick GF (1980) Level dependence of psychophysical frequency resolution and auditory filter shape. J Acoust Soc Am 68:1085–1095.
Pickles JO (1984) Frequency threshold curves and simultaneous masking functions in single fibres of the guinea pig auditory nerve. Hear Res 14:245–256.
Pierce JR, Lipes R, Cheetham C (1977) Uncertainty concerning the direct use of time information in hearing: Place cues in white spectra stimuli. J Acoust Soc Am 61:1609–1621.
Plomp R (1964) The ear as a frequency analyser. J Acoust Soc Am 36:1628–1636.
Plomp R (1967) Pitch of complex tones. J Acoust Soc Am 41:1526–1533.
Plomp R (1976) Aspects of Tone Sensation. London: Academic Press.
Plomp R, Mimpen AM (1968) The ear as a frequency analyzer. II. J Acoust Soc Am 43:764–767.
Plomp R, Steeneken HJM (1968) Interference between two simple tones. J Acoust Soc Am 43:883–884.
Port E (1963) Über die Lautstarke einzelner kurzer Schallimpulse. Acustica 13:212–223.
Ritsma RJ (1962) Existence region of the tonal residue. I. J Acoust Soc Am 34:1224–1229.
Ritsma RJ (1963) Existence region of the tonal residue. II. J Acoust Soc Am 35:1241–1245.
Ritsma RJ (1967) Frequencies dominant in the perception of the pitch of complex sounds. J Acoust Soc Am 42:191–198.
Ritsma RJ (1970) Periodicity detection. In: Plomp R, Smoorenburg GF (eds) Frequency Analysis and Periodicity Detection in Hearing. Leiden, The Netherlands: AW Sijthoff, pp. 250–263.
Scharf B (1961) Complex sounds and critical bands. Psychol Bull 58:205–217.
Scharf B (1970) Critical bands. In Tobias JV (ed) Foundations of Modern Auditory Theory, Volume I. New York: Academic Press, pp. 159–202.
Schooneveldt GP, Moore BCJ (1989) Comodulation masking release as a function of

masker bandwidth, modulator bandwidth and signal duration. J Acoust Soc Am 85:273–281.

Schorer E (1986) Critical modulation frequency based on detection of AM versus FM tones. J Acoust Soc Am 79:1054–1057.

Schouten JF (1970) The residue revisited. In: Plomp R, Smoorenburg GF (eds) Frequency Analysis and Periodicity Detection in Hearing. Leiden, The Netherlands: AW Sijthoff, pp. 41–54.

Schouten JF, Ritsma RJ, Cardozo BL (1962) Pitch of the residue. J Acoust Soc Am 34:1418–1424.

Sellick PM, Russell IJ (1979) Two-tone suppression in cochlear hair cells. Hear Res 1:227–236.

Shailer MJ, Moore BCJ, Glasberg BR, Watson N, Harris S (1990) Auditory filter shapes at 8 and 10 kHz. J Acoust Soc Am 88:141–148.

Shannon RV (1976) Two-tone unmasking and suppression in a forward-masking situation. J Acoust Soc Am 59:1460–1470.

Shepard RN (1964) Circularity in judgments of relative pitch. J Acoust Soc Am 36:2346–2353.

Shower G, Biddulph R (1931) Differential pitch sensitivity of the ear. J Acoust Soc Am 2:275–287.

Siebert WM (1968) Stimulus transformations in the peripheral auditory system. In Kolers PA, Eden M (eds) Recognizing Patterns. Cambridge, MA: MIT Press, pp. 104–133.

Siebert WM (1970) Frequency discrimination in the auditory system: Place or periodicity mechanisms. Proc IEEE 58:723–730.

Small AM (1959) Pure-tone masking. J Acoust Soc Am 31:1619–1625.

Smoorenburg GF (1970) Pitch perception of two-frequency stimuli. J Acoust Soc Am 48:924–941.

Spiegel MF (1981) Thresholds for tones in maskers of various bandwidths and for signals of various bandwidths as a function of signal frequency. J Acoust Soc Am 69:791–795.

Srinivasan R (1971) Auditory critical bandwidth for short duration signals. J Acoust Soc Am 50:616–622.

Srulovicz P, Goldstein JL (1983) A central spectrum model: A synthesis of auditory-nerve timing and place cues in monaural communication of frequency spectrum. J Acoust Soc Am 73:1266–1276.

Stevens SS (1935) The relation of pitch to intensity. J Acoust Soc Am 6:150–154.

Terhardt E (1974) Pitch, consonance and harmony. J Acoust Soc Am 55:1061–1069.

Tyler RS, Hall JW III, Glasberg BR, Moore BCJ, Patterson RD (1984) Auditory filter asymmetry in the hearing impaired. J Acoust Soc Am 76:1363–1368.

Verschuure J (1981) Pulsation patterns and nonlinearity of auditory tuning. II. Analysis of psychophysical results. Acustica 49:296—306.

Verschuure J, van Meeteren AA (1975) The effect of intensity on pitch. Acustica 32:33–44.

Vogten LLM (1974) Pure tone masking: A new result from a new method. In: Zwicker E, Terhardt E (eds) Facts and Models in Hearing. Berlin: Springer-Verlag, pp. 142–155.

Ward WD (1954) Subjective musical pitch. J Acoust Soc Am 26:369–380.

Wegel RL, Lane CE (1924) The auditory masking of one sound by another and its probable relation to the dynamics of the inner ear. Phys Rev 23:266–285.

Wier CC, Jesteadt W, Green DM (1977) Frequency discrimination as a function of frequency and sensation level. J Acoust Soc Am 61:178–184.

Yost WA (1982) The dominance region and ripple noise pitch: A test of the peripheral weighting model. J Acoust Soc Am 72:416–425.

Yost WA, Hill R (1979) Models of the pitch and pitch strength of ripple noise. J Acoust Soc Am 66:400–410.

Yost WA, Hill R, Perez-Falcon T (1978) Pitch and pitch discrimination of broadband signals with rippled power spectra. J Acoust Soc Am 63:1166–1173.

Yost WA, Harder PJ, Dye RH (1987) Complex spectral patterns with interaural differences: Dichotic pitch and the "Central Spectrum." In: Yost WA, Watson CS (eds) Auditory Processing of Complex Sounds. Hillsdale, NJ: Lawrence Erlbaum Associates, pp. 190–201.

Zwicker E (1952) Die Grenzen der Hörbarkeit der Amplitudenmodulation und der Frequenzmodulation eines Tones. Acustica 2:125–133.

Zwicker E (1956) Die elementaren Grundlagen zur Bestimmung der Informationskapazität des Gehörs. Acustica 6:365–381.

Zwicker E (1961) Subdivision of the audible frequency range into critical bands (Frequenzgruppen). J Acoust Soc Am 33:248.

Zwicker E (1965a) Temporal effects in simultaneous masking by white-noise bursts. J Acoust Soc Am 37:653–663.

Zwicker E (1965b) Temporal effects in simultaneous masking and loudness. J Acoust Soc Am 38:132–141.

Zwicker E (1970) Masking and psychological excitation as consequences of the ear's frequency analysis. In Plomp R, Smoorenburg GF (eds) Frequency Analysis and Periodicity Detection in Hearing. Leiden, The Netherlands: AW Sijthoff, pp. 376–394.

Zwicker E, Fastl H (1972) On the development of the critical band. J Acoust Soc Am 52:699–702.

Zwicker E, Feldtkeller R (1967) Das Ohr als Nachtrichtenempfänger. Stuttgart: Hirzel.

Zwicker E, Scharf B (1965) A model of loudness summation. Psychol Rev 72:3–26.

Zwicker E, Flottorp G, Stevens SS (1957) Critical bandwidth in loudness summation. J Acoust Soc Am 29:548–557.

4
Time Analysis

NEAL F. VIEMEISTER AND CHRISTOPHER J. PLACK

1. Introduction

In audition, perhaps more than in any other sense, the temporal aspects of the stimulus are crucially important for conveying information. This clearly is true of speech and of most auditory communication signals, where the temporal pattern of spectral changes is, essentially, the informational substrate. Indeed, an auditory "pattern" is seldom a fixed spectral shape; rather, it is a time varying sequence of spectral shapes. Given the fundamental importance of temporal changes in audition, it is not surprising that most auditory systems are "fast," at least compared to other sensory systems. We can hear temporal changes in the low millisecond range. We can, for example, hear the roughness produced by periodically interrupting a broadband noise at interruption rates up to several kHz. This is several orders of magnitude faster than in vision where the analogous "flicker fusion frequency" is a sluggish 50 to 60 Hz.

This chapter reviews and highlights the psychophysics of auditory temporal processing. The major focus is on temporal resolution, the ability to follow rapid changes over time. Also considered are temporal aspects of masking, particularly forward masking and overshoot, and temporal integration, the ability to combine information over time.

2. Temporal Characteristics of Sound: Envelope and Fine Structure

Sound consists of rapid pressure changes in an acoustic medium; temporal changes are, therefore, inherent in the nature of sound. However, for the purposes of this chapter, it is necessary to distinguish between the rapid pressure variations that carry the acoustic information (the "fine structure") from the slower, overall changes in the amplitude of these fluctuations (the

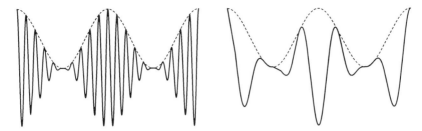

FIGURE 4.1. Sinusoidal amplitude modulation with $m = 1$ and $f_c = 10f_m$ (left panel) or $f_c = 2.5f_m$ (right). The dashed line is the modulation waveform.

"envelope"). Temporal resolution, for example, normally refers to the resolution of changes in the envelope of a sound, not in the fine structure.

The notion of an envelope can be somewhat elusive and bears brief discussion (see Rice 1982). Consider the signal $s(t) = a(t)\cos[2\pi f_c t + \varphi(t)]$. The "instantaneous amplitude" of $s(t)$ is $a(t)$ and its magnitude, $|a(t)|$, is the "envelope."[1] When $a(t)$ and $\varphi(t)$ are "slowly varying" relative to the carrier frequency, f_c, $s(t)$ is narrow band, and the meaning of the envelope is clear and corresponds to the intuitive notion of the envelope as the boundary of the waveform. When $a(t)$ and $\varphi(t)$ are not slowly varying, specifically, when they have spectral components that are more than about $f_c/2$, the intuition is lost. Figure 4.1 illustrates this for the specific case of a sinusoidally amplitude-modulated (SAM) tone, $s(t) = [1 + m\cos(2\pi f_m t)]\cos(2\pi f_c t)$, where f_m is the modulation frequency and m is the modulation index or modulation depth ($0 \le m \le 1$). In the left panel, $f_c = 10f_m$, and the boundary of the waveform is well described by the modulating waveform, $[1 + m\cos(2\pi f_m t)]$, shown by the dashed lines in Figure 4.1. In the right panel, however, $f_c = 2.5f_m$, and the fine structure does not exhibit a clear boundary or envelope. The issue here is what is meant by "slowly varying amplitude" in defining the envelope. It is closely related to the issue of what is meant by a "narrowband" signal; in a sense, only narrowband signals have envelopes, although it is convenient to discuss the envelope of broadband waveforms such as speech and amplitude-modulated noise.

A useful definition of the envelope is based on the concept of an analytic signal. The analytic signal corresponding to $s(t)$ is the complex signal that has $s(t)$ as its real part and the Hilbert transform of $s(t)$ as its imaginary part (see Panter 1965). The envelope of $s(t)$ is the magnitude of the analytic signal. Although techniques based upon this definition are frequently used for extracting the envelope of a waveform, it should be noted that the "envelope" so obtained may not be slowly varying with respect to the fine structure.

---

1. The "instantaneous frequency" is usually defined as the time derivative of the instantaneous phase, $2\pi f_c t + \varphi(t)$, and is $f_c + \frac{d}{dt}\varphi(t)$, in Hz.

In Figure 4.1, for example, the envelopes obtained via the analytic signals are both $[1 + m\cos(2\pi f_m t)]$, even though there is no clear boundary for the signal shown in the right panel.

The "envelope detector" provides an alternative definition, one that seems more useful for audition. In fact, envelope detectors have been used extensively in modeling the initial stages of auditory processing (Jeffress 1967). Some of these models will be described in Section 5. An envelope detector consists of a nonlinearity, typically half-wave rectification, followed by a low-pass filter or "leaky integrator." For the SAM tone discussed above, the nonlinearity introduces a DC component, a component whose frequency is f_m and components at higher frequencies. The low-pass filter attenuates the higher frequency components leaving, essentially, a waveform that can closely approximate the envelope. The characteristics of the low-pass filter, its cutoff frequency and attenuation rate, determine the amount of fine structure allowed in this waveform. Essentially, by specifying these characteristics, one is also specifying what is meant by "slowly varying."

3. Temporal Resolution: General Considerations

3.1 Definition of Temporal Resolution

Temporal resolution or temporal acuity refers to the ability of the auditory system to respond to rapid changes in the envelope of a sound over time. If the response of the system is sluggish (i.e., temporal resolution is poor), then the internal representation of the sound will not faithfully represent the temporal changes present in the physical stimulus. Information about these changes has been lost at some stage. In this case, we can imagine the representation of the sound being "blurred" in time in a similar way as an out-of-focus visual image is blurred in space.

In this chapter, two aspects of temporal resolution will be considered: "with-in-channel" and "across-channel." Within-channel resolution refers to changes that occur within a relatively limited frequency region, i.e., a single critical band. This is contrasted with the "across-channel" resolution of changes that occur between critical bands. The focus will be on within-channel resolution. This emphasis reflects the far more extensive research in this area and our better understanding of the processes involved (see also Yost and Shelf, Chapter 6).

3.2 Why is Temporal Resolution Limited?

It is convenient to distinguish two general sources of temporal resolution limitation in the auditory system: peripheral limitations and central limitations. (The term "peripheral" is used to refer to the first stages of auditory processing, up to and including the auditory nerve). The concept of a central limitation is somewhat vague since all the aspects of auditory processing about which very little is known are effectively lumped together. It remains a useful concept, however, if only as a complement to the concept of a peripheral limitation for which there is a considerable amount of relevant physiological and psychophysical data.

3.2.1 Peripheral Limitations

The transduction process in the cochlea cannot maintain perfect temporal acuity. First, the filtering action of the basilar membrane necessarily limits temporal resolution by restricting the listening bandwidth. As a rule of thumb, temporal fluctuations at a higher frequency than the bandwidth of the filter will not be passed by the filter. The effect of filtering is likely to dominate at low frequencies, where the critical bandwidth is small and hence the temporal response, or ringing, of the auditory filter is long. The evidence for the role of the auditory filters in temporal resolution will be considered in Section 6.1.

The properties of hair cells, synapses, and the refractory period of neurons limit the firing rate that can be achieved in the auditory nerve. This will impose a limit on the rate of envelope fluctuations that can be encoded (although a higher effective firing rate, and hence superior temporal resolution, might be realized by combining information across neurons). In addition, at low frequencies, it is possible that phase locking may impose a limit on resolution: if a nerve fiber is only firing at a particular phase in the period of a 100-Hz tone, for example, the time between spikes or bursts of spikes will be at least 10 ms. This will be the lower limit on acuity at this frequency.

3.2.2 Central Limitations

Central limitations may result from the processing of information from the auditory nerve fibers. Individual nerve impulses are "all-or-nothing," i.e., information is not encoded in the magnitude of the spike; information about sounds can only be encoded as the temporal pattern of neural firings. This means that the auditory system has to have some way of combining information over time, for example, by computing a mean firing rate or a mean interspike interval for the neuron. This processing will necessarily result in a reduction in temporal resolution. To follow shallow, rapid fluctuations in the envelope of a sound, for example, it is necessary to estimate intensity accurately over a succession of brief time periods. It would appear, therefore, that some type of integration is necessary, although the actual duration of the integration may not need to be large if information from a group of fibers is combined.

The hypotheses described in this section are almost purely speculative; it has not been demonstrated conclusively, for example, that a central limitation to resolution exists. It appears likely, however, that high-level auditory processing will produce a certain degree of sluggishness in the system, simply because of the constraints imposed by neural encoding.

4. Simple Estimates of Within-Channel Acuity

Most of the research in this area in the past has been concerned with describing temporal resolution by a single number: the smallest time interval necessary for a change to be detected. Unfortunately, there is a difficulty associated with trying to measure the detectability of temporal changes; any change in the temporal characteristics of a sound necessarily affects the magnitude and/or the phase spectrum of the sound. In some cases, listeners are able to use spectral changes in order to detect the temporal change. For example, Leshowitz (1971) has shown that the minimum gap between two clicks required to distinguish them from a single click of equal overall energy is 6 μs, suggesting remarkably fine temporal resolution. However, this discrimination was almost certainly performed using spectral information (in this case, slight differences in the stimulus energy at high frequencies). Experiments on temporal resolution have to be designed to minimize the spectral cues that are available to the listener so that discrimination can only be achieved by the detection of envelope changes. There are two ways of doing this. One way is to change the time envelope of the stimuli using phase changes which don't affect the long-term magnitude spectrum of the sound. The second technique is to mask the spectral changes that occur so that they cannot be detected. Examples of each of these two approaches are presented in Sections 4 and 5.

4.1 Time-Reversed Stimuli

Ronken (1970) asked subjects to discriminate between two pairs of clicks. Each click had a duration of 250 μs. In one pair, the first click was higher in amplitude than the second click and conversely for the second pair. These two pairs have identical long-term magnitude spectra, but differ in their phase spectra. By varying the time interval between the clicks in each pair, Ronken found that these pairs could be discriminated down to a click separation of around 2 ms. Henning and Gaskell (1981), on the other hand, using clicks only 20 μs in duration, showed that in some cases click pairs could be discriminated when the interval between the clicks was only 200 μs. The reason for this discrepancy is not clear.

Green (1973) conducted an experiment similar to that of Ronken (1970) except that he used brief pulses of temporally contiguous sinusoids instead of clicks. The task was to differentiate a stimulus where the first sinusoid in a pair was more intense than the second from the case where the second was more intense. The threshold in this task was between 1 and 2 ms for the total duration of each stimulus.

4.2 Gap Detection

Probably the most frequently used technique for measuring temporal resolution has been the gap detection experiment. Subjects are required to detect a brief temporal gap in a band of noise or a sinusoid. The sounds before and after the gap are often referred to as "markers." For broadband markers, introducing an abrupt temporal gap into the stimulus does not change the magnitude spectrum, and, hence, spectral changes cannot be used as a cue to the presence of the gap. The threshold gap duration in this case is around 2 to 3 ms (Plomp 1964; Penner 1977). When sinusoidal or bandpass noise markers are used, the gap generates spectral cues which have to be removed in some way, for example, by masking with a simultaneous notched noise. Sinusoidal markers are particularly problematic because it is difficult to design a masking stimulus that masks splatter but does not interfere with the detectability of the gap. When spectral cues have been reduced, the gap threshold has typically been measured as 2 to 3 ms for noise band markers (Fitzgibbons and Wightman 1982; Fitzgibbons 1983; Shailer and Moore 1983; Buus and Florentine 1985) and 4 to 5 ms for sinusoidal markers (Shailer and Moore 1987). These estimates of acuity are broadly similar to those from the click pairs studies.

5. The Systems Analysis Approach

The results discussed so far seem to suggest that there is a lower limit on acuity of about 3 ms. Describing temporal resolution by a single number, the smallest detectable gap or the shortest duration of a click pair necessary for discrimination, is a limited approach, however. To understand the reason for this, it is first necessary to describe how researchers have attempted to model temporal resolution.

Essentially only one type of model has been developed for describing temporal resolution. The model is based on a linear systems analysis approach (Rodenburg 1977; Viemeister 1977, 1979) and is similar to the envelope detector described in Section 2. This approach was initially used to describe temporal and spatial processing in vision, and the early work in audition (Viemeister 1970) was inspired by its successful application in vision. The basic model consists of four stages: (i) a bandpass filter; (ii) a nonlinearity; (iii) a low-pass filter or leaky integrator; and (iv) a decision mechanism. The differences between the models arise from the different properties assigned to the four stages. The bandpass filter corresponds to the peripheral auditory filter and its characteristics change with center frequency. The nonlinear stage has two possible functions: first, to introduce low-frequency components corresponding to the envelope of the signal (for this, simple rectification is often used, e.g., Viemeister 1979), and, second, to simulate nonlinearities in the auditory system. Penner (1980) has suggested that the nonlinear stage should be highly compressive to account for nonlinearities in the properties of nonsimultaneous masking, such as the nonadditivity of forward and backward masking. Approaches based on energy integrators or temporal windows usually implicitly assume that the nonlinearity is a square-law, an expansive nonlinearity. The next stage of low-pass filtering/integration simulates the temporal resolution limitation by removing rapid changes in the envelope of the signal. The characteristic of the leaky integrator is often assumed to be given by the modulation transfer function (see Section 5.1) or the temporal window (see Section 5.2), although the form of these functions usually depends on all stages of the system because of the measurement techniques employed. The decision mechanism is intended to simulate how a subject uses the output of the leaky integrator to make a discrimination in a particular task. The mechanism might use the signal-to-noise ratio at a particular time in the stimulus (Moore et al. 1988), the overall variance of the output of the leaky integrator (for detecting modulation, Viemeister 1979), or the ratio between the maximum and minimum values of the output (Forrest and Green 1987).

To demonstrate the operation of this model, the left panel of Figure 4.2 shows the output of each of the first three stages of a representative version of the model in response to a 100-ms burst of wideband noise (0 to 10 kHz) amplitude modulated at a rate of 50 Hz with a 100% modulation depth. The top row shows the input waveform. The modulation can be seen quite clearly. The next row down shows the output of a simulated auditory filter with a center frequency of 2 kHz in response to this input. The modulation is less distinct here because modulation components distant from 2 kHz have

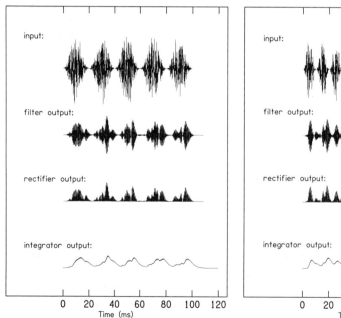
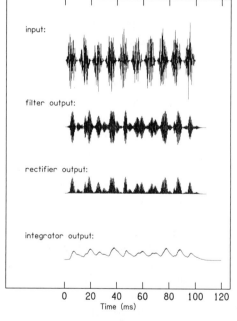

FIGURE 4.2. The output of the various stages of the model described in the text in response to a 50-Hz SAM wideband noise (left panel) and a 100-Hz SAM wideband noise (right panel).

been attenuated by the filter. Next is the output of the half-wave rectifier stage and, finally, at the bottom is the output of a leaky integrator with a time constant of 3 ms. The depth of modulation at the output of the integrator is less than that of the input waveform. In other words, the temporal changes in the input are not resolved perfectly by the model. The right panel of Figure 2 shows the model's response to 100-Hz amplitude modulation. The modulation depth at the output of the integrator is even less in this case. If the modulation frequency were to increase systematically, the modulation depth of the modulator frequency component at the output of the integrator would decrease until it was no longer discriminable from random fluctuations in the noise. The model has the effect of attenuating rapid envelope fluctuations.

In light of this approach, it can now be understood why single-value measures of acuity are not adequate. Consider, for example, the gap detection task. Assuming the basic form of the model outlined above is a reasonably accurate description of temporal processing, then the internal representation of the gap (the output of the leaky integrator) will be a smooth dip. The shorter the gap, the shallower the dip will be. The gap detection task can therefore be seen to involve two components: a temporal resolution component that determines the depth of the dip and an intensity discrimination component that determines the smallest dip that can be detected. In other words, the gap detection experiment cannot be regarded as a measurement of temporal resolution per se. Using a task that required subjects to detect brief decrements in the intensity of a wideband noise (rather than a period of complete silence, as in the gap detection experiments described above), Buunen and van Valkenburg (1979) showed that the gap threshold is dependent on the depth of the decrement. The smaller the change in intensity, the larger the gap threshold. This illustrates the importance of intensity differences in gap detection. The same basic argument also applies to the time-reversed stimuli. In this case, the temporal resolution of the system determines the difference in the envelopes of the two click pairs while the intensity resolution of the system determines whether these differences will be detected.

An alternative approach to measuring temporal resolution is based on the framework described earlier in this section. This approach assumes that the primary resolution limitation results from a low-pass filter or leaky integrator; rapid changes are removed and slow changes are unaffected. Measuring the characteristics of the leaky integrator should give a description of temporal resolution that is independent of the efficiency of the auditory system (its ability to detect small changes in intensity) and that can predict discrimination based both on the duration of changes in the envelope and the magnitude of the changes (the "contrast"). Two experimental approaches based on this understanding will be described in Sections 5.1 and 5.2.

5.1 Temporal Modulation Transfer Functions

The notion underlying the use of temporal modulation transfer functions (TMTFs) is that many aspects of temporal processing can be described by a transfer function that describes how sinusoidal envelopes of different frequencies are attenuated and phase shifted by the auditory system. This transfer function is often assumed to be equivalent to the leaky integrator stage of the model just described. The appeal of this approach is that if the system is linear, "envelope linear" for the TMTF, then the transfer function permits determination of the effective output envelope for arbitrary envelopes of the input. Envelope linearity means, essentially, that the output of the "black box" in response to a SAM waveform is a sinusoid at the modulation frequency plus a DC component, i.e. a linear transformation of the envelope of the input. It is almost certainly true that the auditory system is not envelope linear for large envelope excursions; thus, considerable caution is necessary in applying TMTFs as a general description of temporal processing.

5.1.1 Measurement of TMTFs

Psychophysical TMTFs are most frequently based upon modulation thresholds for SAM noise. A modulation threshold is the modulation depth (m) required to just discriminate between the SAM carrier ($m \neq 0$) and the unmodulated carrier. The relationship between the modulation threshold and modulation frequency is typically called the TMTF although, ideally, it is only its attenuation characteristic. The basic assumption is that, at the modulation threshold, the amplitude of the output of the envelope-linear black box is constant, independent of modulation frequency. If this is true, then the sensitivity to modulation ($1/m$) is directly related to the attenuation of the envelope by the system.

Another approach involves measurement of the threshold for a brief probe at various temporal locations within a period of a SAM masker. Consistent with envelope linearity, such masking patterns are usually sinusoidal (Viemeister 1977). The amplitude and phase of the best fitting sinusoids for masking patterns measured at different modulation frequencies can be used to obtain an amplitude and a phase characteristic of the TMTF. The phase characteristic typically is flat, except perhaps at high modulation frequencies. The attenuation characteristic is a low-pass function but appears to roll off at a greater attenuation rate than that seen using modulation thresholds. This steeper rolloff may result from the use of finite duration probes: as modulation frequency increases, the probe occupies an increasingly large portion of the modulation cycle, thus "smearing" the masking pattern. It may be possible to correct for this smearing, but there remains the more general problem of justifying the assumption underlying this technique that the probe threshold is proportional to the instantaneous output of the temporal processor. Although the probe procedure offers several advantages over other procedures, the general problem of smearing seems to preclude its usefulness for studying temporal resolution. Surprisingly, this problem has not been noted in studies of the critical masking interval or in the measurement of temporal windows (see Section 5.2). A similar procedure has been used for other purposes in the context of "masking period patterns" (Zwicker 1976).

TMTFs have been obtained using other procedures (see Viemeister 1979), but these methods have either limitations or assumptions more debilitating than the methods described above. By far, most empirical TMTFs have been obtained using modulation thresholds; the remainder of Section 5.1 will focus on them and on some of the underlying issues.

5.1.2 General Characteristics and Issues

Sinusoidal carriers would appear to be ideal for using TMTFs to study within-channel temporal resolution. Unfortunately they are not, except perhaps at very high carrier frequencies. The problem is that, at relatively moderate modulation frequencies, e.g., 60 to 100 Hz for 1-kHz carriers, the sidebands produced by modulation can be resolved by the peripheral auditory system and the modulation is detected spectrally, not temporally as intended. This limitation is reduced as the carrier frequency is increased because of the decrease in frequency resolution with increasing center frequency. Nevertheless, even for high carrier frequencies, spectral cues remain a serious limitation.

A solution to the problem of spectral cues is to use broadband noise as a carrier. The long-term spectrum of SAM noise is flat and invariant with changes in modulation frequency. An issue is whether the short-term spectrum of SAM noise is also flat (Pierce, Lipes, and Cheetham 1977) and, more directly, whether the modulation might be detected by momentary changes in the shape of the spectrum. This issue appears repeatedly in discussions of the pitch of SAM noise (see Burns and Viemeister 1981) and is also relevant here, especially at higher modulation frequencies. Briefly, in SAM noise, the rapid changes in the spectral density at frequency f_c are correlated with changes at frequencies $f_c - f_m$ and $f_c + f_m$, where f_m is the modulation frequency. A momentary peak at f_c, for example, also tends to cause peaks at these other frequencies. At high modulation frequencies where these peaks can be resolved, it is conceivable that modulation is detected based upon correlated changes in momentary spectral peaks. It is not clear, however, that under realistic constraints on spectral and temporal resolution these cues are useful to the auditory system. Furthermore, modulation thresholds are unaffected or improve when modulation information is restricted to high frequencies (Viemeister 1979), frequencies for which spectral resolution is relatively poor. This, of course, suggests that short-term spectral cues are not being used.

A typical TMTF for broadband noise is shown in Figure 4.3. To preserve a sense similar to that for an attenuation characteristic, modulation thresholds increase downward on the y-axis. This TMTF shows a low-pass characteristic with a -3 dB cutoff at 50 Hz and an asymptotic attenuation rate of approximately 4 dB/octave. At very high modulation frequencies where the modulation threshold is large, the introduction of modulation produces an

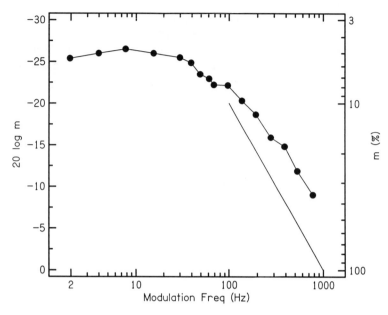

FIGURE 4.3. TMTF obtained with a continuous broadband noise carrier at 30 dB spectrum level. Each point is the modulation threshold averaged over four subjects with normal hearing. The solid line has an attenuation rate of 6 dB/octave. (Data are from Bacon and Viemeister 1985.)

increase in average intensity that subjects apparently use to "detect" the modulation; the TMTF shows a high-frequency asymptote of approximately -7 dB for modulation frequencies above 1 kHz. The data shown in Figure 4.3 were obtained under conditions for which this intensity increase was compensated for by reducing the level of the carrier, thus presumably eliminating this cue. An alternative, possibly preferable approach, is to "rove" the carrier level (Forrest and Green 1987). This yields similar results, namely, an extension of the 4 dB/octave slope to very high modulation frequencies. Extrapolation of the TMTF of Figure 4.3 to 100% modulation yields an upper cutoff modulation frequency of approximately 2.5 kHz, a value close to that measured more directly by Viemeister (1979). This is a surprisingly high frequency and is close to the putative limit of phase locking.

The low-pass characteristic of the TMTF, such as that shown in Figure 4.3, is generally observed for continuous carriers and durations of modulation longer than 250 ms. For shorter durations, there is a tendency for modulation thresholds to increase more at low modulation frequencies. This is especially true with gated carriers where, for durations less than 500 ms, a high-pass segment appears and the TMTF shows a bandpass characteristic. The increased thresholds for low modulation frequencies at relatively brief durations appear to reflect, in part, a reduction in the number of "looks" at the envelope fluctuations (Viemeister 1979; Sheft and Yost 1990). This, however, does not explain the appearance of a high-pass segment with gated carriers. It has been suggested that the high-pass segment may result from adaptation (Viemeister 1979). Sheft and Yost (1990) have shown, however, that the high-pass segment is present even when the carrier is gated 500 ms before the modulation. This seems excessively long to be explained by adaptation.

5.1.3 Theoretical Considerations

The general characteristics of the TMTF are roughly consistent with the leaky integrator in the model discussed in the introduction to Section 5. Indeed, many investigators simply fit the TMTF with the attenuation characteristic of a simple low-pass filter to estimate a time constant. This may be adequate for certain purposes, but it clearly is an oversimplification and may obscure some important theoretical issues.

The initial stages of Viemeister's model (1979) consisted of the familiar band-pass filter, followed by a half-wave rectifier, followed in turn by a simple low-pass filter. The decision statistic is the root-mcan-squared (rms) value of the output of the low-pass filter computed over the duration of the observation interval. This statistic was chosen because it seems fairly realistic and offers some basis for describing other phenomena such as the pitch of SAM noise and discrimination of modulation frequency. With a filter bandwidth of 2 kHz and a low-pass cutoff of 65 Hz (2.5 ms), the predicted TMTF is in good agreement with data obtained without intensity compensation. As Forrest and Green (1987) have noted, this model predicts a 6 dB/octave attenuation rate when intensity compensation or roving is used; this is significantly greater than the 4 dB/octave typically observed. One approach is to abandon the simple low-pass filter with its 6 dB/octave roll-off and simply assume that the low-pass filter has an attenuation rate of 4 dB/octave. Forrest and Green retained the simple low-pass filter and proposed an alternative decision statistic, the ratio of the largest to the smallest instantaneous output of the low-pass filter. Simulating modulation detection using a 4-kHz filter and a simple low-pass filter with a time constant of 3 ms, they showed that this "max/min" decision statistic predicts an intensity-compensated TMTF in good agreement with the data. They also show that this same model provides a good account of the detection of partially filled gaps in noise. Sheft and Yost (1990) argued, however, that the rms statistic provides a better account than the max/min statistic of the change in modulation threshold with duration. They also showed that when the half-wave rectifier is replaced with a neural transduction model that includes adaptation, the model better describes the bandpass TMTF observed with gated carriers. It does not appear, however, that this adaptation will account for the finding, discussed in Section 5.1.2 that a 500-ms fringe does not eliminate the high-pass segment.

The nature of the nonlinearity and the decision statistic used for modulation detection are important outstanding issues. A more fundamental issue, however, is the bandwidth of the initial bandpass filter that is required by these models. This bandwidth is 2 to 4 kHz, nearly an order of magnitude larger than the typical critical bandwidth estimates. A wide bandwidth is required to account for the detectability of modulation at high modulation frequencies. If a typical auditory filter characteristic or a typical psychophysical tuning curve is used for the initial filter, these models will grossly overestimate the modulation thresholds at high modulation frequencies.

If, as seems likely, peripheral filtering does not determine the form of the TMTF (see Section 6.1), we are faced with the question of how the "system" overcomes the bandwidth limitations of the peripheral filters. One can, of course, increase the bandwidth by combining the outputs of filters tuned to different frequencies. To be realistic, this must be done by pooling spikes across different fibers. It is not clear whether such a pooling scheme will permit reasonably faithful recovery of the envelope.

5.1.4 Additional Issues

There are several interesting phenomena involving modulation detection that complicate the rather simple picture suggested by the models used to describe the TMTF. Modulation detection interference (MDI) is a term coined by Yost and Sheft (1989) to describe the elevation in modulation threshold of a sinusoidal carrier that occurs when modulation at the same rate is present at a different carrier frequency (see Green, Chapter 2; Yost and Sheft, Chapter 6). They showed, for example, that the threshold for detecting 10-Hz SAM modulation of a 1-kHz carrier increases by 20 dB ($20 \log m$) when a 4-kHz carrier with 10 Hz, 100% modulation is present. Yost and Sheft viewed MDI as another manifestation of a "cross-spectral" grouping process that occurs when different frequency regions share common modulation. Somewhat troublesome for this account is the fact that changes in the phase relationship between the modulators seem to have an inconsistent effect; some subjects show large MDI with no phase effect, others show large phase effects. Also troublesome is that some MDI is seen when the carriers are presented dichotically (Yost and Sheft 1990). Changes in phase or dichotic presentation might be expected to affect the "grouping" and therefore produce changes or, with dichotic presentation, eliminate MDI.

The fact that MDI is seen both with wide frequency spacing between the carriers and dichotically suggests that it is a cross-channel effect. Caution should be exercised, however, in concluding that wide frequency spacing implies cross-channel processing. Wakefield and Viemeister (1985) showed, for example, that a 200-Hz tone can markedly affect the detectability of 200-Hz modulation of a band of noise centered at 10 kHz. This occurs even when the 200-Hz tone produces no masking at 10 kHz and over a wide range of carrier levels. Although this phenomenon appears to be different from MDI (see Yost and Sheft 1989), it does indicate that substantial, and mysterious, interactions, presumably within-channel, can occur over extremely large frequency separations.

It has recently been shown, using broadband noise carriers, that the presence of AM at a given modulation frequency can affect the detectability of AM at other modulation frequencies. Bacon and Grantham (1989) measured "modulation masking patterns" and showed, for example, that 64 Hz, 50% modulation increased the modulation thresholds for signal modulation frequencies from 8 to 256 Hz, with most masking (10 dB) occurring, as expected, at 64 Hz. Their results, results from a somewhat similar experiment by Houtgast (1989), and unpublished data from our laboratory indicated that there is some degree of frequency selectivity for modulation frequency but, at least in comparison with "spectral" frequency selectivity, it is rather poor. The masking patterns obtained by Bacon and Grantham, for example, show a -10 dB "bandwidth" of two to three octaves. Yost, Sheft, and Opie (1989) also found poor modulation frequency selectivity in MDI with 1- and 4-kHz carriers.

Modulation depth discrimination is a special case of modulation masking in which the signal and masker modulation frequencies are the same. It is not clear what an appropriate measure of the just noticeable difference (JND) is for modulation depth but, when the difference in modulation *power* is used, analogous to ΔI, the JND increases for modulation depths of the standard greater than 5 dB above modulation threshold. The increase in the JND is not, however, consistent with Weber's Law (Wakefield and Viemeister 1990).

A perplexing issue in modulation detection concerns the possible role of adaptation to AM. It has been demonstrated that exposure to a SAM waveform at a given modulation frequency will selectively elevate modulation thresholds for modulation frequencies near the modulation frequency of the adaptor (Tansley and Suffield 1983). This presents a potential problem for the studies described above in which large modulation depths are used for the "masker." Wakefield and Viemeister (1990) noted, for example, strong sequential dependencies in threshold estimates across conditions: thresholds were substantially higher when large modulation depths had been used on the previous block. It is not clear how such adaptation might have affected the results on MDI and modulation masking, but it is conceivable that adaptation obscured phase effects in MDI or produced unusually broad modulation masking patterns.

Selective adaptation has been taken as evidence for "channels" tuned to specific stimulus properties such as AM (Green and Kay 1973) or FM (Green and Kay 1974). This evidence is not particularly compelling: the adaptation effects typically are small, they may disappear with training (Moody et al. 1984), and they may reflect nonsensory factors (Wakefield and Viemeister 1984). Nevertheless, Bacon and Grantham (1989) invoked AM channels to explain the tuning indicated by the modulation masking patterns they obtained. Similarly, Yost, Sheft, and Opie (1989) employed AM channels to explain MDI and, more generally, to account, at least roughly, for the formation of auditory "objects" based upon common modulation. Yost and Sheft (1989) cautiously pointed to recent physiological data that suggested "tuning" to AM by units in the inferior colliculus (Rees and Møller 1983) and in the cortex (Schreiner and Urbas 1988). The relevance of the physiological data to the psychophysical findings is, of course, uncertain. Although postulating AM channels offers some superficial appeal, it is our opinion that, without stronger evidence supporting their existence, such a modeling strategy is premature and may be misleading (see also Yost and Sheft, Chapter 6).

5.2 The Temporal Window

A different approach from the empirical (but not theoretical) point of view is to measure directly the shape of the hypothetical leaky integrator in the model. This takes the form of a weighting function that is assumed to smooth the internal representation of auditory stimuli by integrating energy over time. This process is comparable to performing a running average of the incoming stimulus; the result of the processing is the output of the temporal window as a function of time.

It can be argued that it is easier conceptually to describe resolution in terms of an integration in the time domain rather than as a transform in the frequency domain, such as the TMTF, although the two approaches are equivalent; ideally, the temporal window should be the mirror image of the impulse response of the TMTF.

5.2.1 The Critical Masking Interval

Experiments conducted to measure the characteristics of the temporal window have, in many respects, paralleled experiments that have measured the characteristics of the auditory filter, with time substituted for frequency as the dimension of interest (see Moore, Chapter 3). In the early 1970s, Penner and her colleagues (Penner, Robinson, and Green 1972; Penner and Cudahy 1973) introduced the idea of the critical masking interval (CMI) as the temporal analogue of the critical band description of frequency selectivity. By measuring the threshold of a click as a function of the duration of a noise burst temporally centered on the click, they showed that, for noise durations greater than between 6 and 30 ms, increasing the duration of the noise had little effect on threshold (cf. Fletcher's 1940 classic experiment on the critical band). This duration is the CMI, which can be regarded as the "effective" duration of the temporal window, just as the critical bandwidth is a simple estimate of frequency resolution. It is very difficult, however, to get an accurate measure of the CMI using this technique (hence, the wide variability in estimates of the CMI). Furthermore, temporal masking effects such as overshoot (see Section 8.2) complicate the interpretation of the results.

5.2.2 The Shape of the Temporal Window

Since the work on the CMI, masking studies have attempted to determine the shape of the window more precisely. Festen and colleagues (Festen et al. 1977; Festen and Plomp 1981,1983; Festen 1987) measured window shapes using a method similar to the probe technique employed by Viemeister (1977) and described in Section 5.1.1. They measured the threshold for filtered clicks presented at a peak or in a valley of sinusiodally intensity modulated (SIM) noise as a function of the modulation rate. Depending on the duration and shape of the window, high-frequency modulation is smoothed more than low-frequency modulation and, therefore, the difference between the thresholds for the peak and the valley placements tends to decrease as the modulation rate is increased. If the peak and valley thresholds are measured for a range of modulation rates, the shape of the temporal window can be estimated.

One limitation of the SIM noise experiments is that, because the masking stimulus is symmetrical, they do not give any information about the possible asymmetry of the temporal window. Robinson and Pollack (1973) suggested that the slope of the window was shallower for times before the center than for times after on the basis that forward masking is more temporally extensive than backward masking. To examine this question, Moore and colleagues (Moore et al. 1988; Plack and Moore 1990) derived window shapes by measuring the threshold of a tone pulse positioned in a gap between two bursts of noise. This is a temporal analogue of the technique first described by Patterson (1976) for measuring the shape of the auditory filter. In the temporal case, the threshold of the tone depended on combined forward and backward masking. By systematically varying both the duration of the gap and the temporal position of the tone within the gap, it was possible to estimate the overall form and asymmetry of the window. The analysis used assumed that threshold was determined by the maximum signal-to-noise ratio obtainable at the output of the temporal window. As expected, the windows were asymmetric in the direction suggested by Robinson and Pollack. The windows they derived had time constants comparable to a CMI of about 8 ms.

A potential problem with the forward and backward masking technique concerns the effect of adaptation on the results. It is assumed that the masking produced by the noise bursts is due to the temporal smearing effect of the leaky integrator, effectively producing a simultaneous masking situation. If threshold is largely determined by neural adaptation due to the forward masker, however, then the technique is not measuring the shape of the hypothetical temporal integrator but rather the characteristics of the recovery from adaptation. The role of adaptation in forward masking will be discussed in greater detail in Section 8.8.1.

Plack and Moore (1991) showed how the shape of the temporal window could also be derived from data on the detection of decrements in broadband noise. The temporal windows they derived were roughly comparable to those from the masking study.

6. Parametric Effects

6.1 Temporal Resolution and the Auditory Filter

As discussed by Moore (Chapter 3), the bandwidth of the bandpass auditory filters in the cochlea decreases as the center frequency of the filters decreases. As the bandwidth of a filter is reduced, the temporal response or "ringing" of the filter is prolonged, so that the filter will continue to oscillate for a longer time after stimulation has ceased. The low-frequency filters in the cochlea, therefore, have a longer temporal response than the high-frequency filters. This result has been clearly demonstrated in the mammalian auditory system by physiological techniques. For example, in the cat cochlea, the neural response to a click stimulus in an auditory nerve fiber with a center frequency of 8 kHz lasts approximately 2 ms, whereas the response of a fiber tuned to 1 kHz lasts 8 ms (Kiang et al. 1965). It has been suggested that this ringing response is responsible for limiting temporal resolution in the auditory system and, consequently, resolution should deteriorate with decreasing center frequency (Shailer and Moore 1983; Buus and Florentine 1985). At high frequencies, the temporal response of the auditory filter is very brief and it might be suspected that performance is dominated by a more central component (i.e., the low-pass filter in the model discussed in Section 5). At low frequencies, however, the effect of the auditory filter may be considerable.

6.1.1 The Variation in Temporal Resolution with Center Frequency

Experiments using narrowband noise markers have frequently shown a decrease in gap threshold as the center frequency of the noise band is increased (Tyler et al. 1982; Fitzgibbons 1983; Shailer and Moore 1983, 1985; Buus and Florentine 1985). The use of noise markers is problematic, however. A noise band contains inherent random envelope fluctuations. The narrower the bandwidth of the noise, the more these fluctuations are dominated by low-frequency components. Even if the physical bandwidth of the noise is large, the effect of the auditory filter is to filter out a narrow band of noise. The lower the center frequency of the filter, the narrower the bandwidth of the filtered noise becomes. It is probable that the subjects' gap detection performance at low frequencies with noise markers is limited, not by temporal resolution per se but by their ability to discriminate the gap from dips in the envelope of the noise (Green 1985; Shailer and Moore 1985). Shailer and Moore (1985) found a considerable effect of bandwidth on gap detection in narrowband noise. For a 50-Hz bandwidth, gap thresholds could be as high as 45 ms. The bandwidth problem might account for most of the frequency effect seen in experiments using noise markers. Gap thresholds measured with sinusoidal markers (which have flat envelopes, even after filtering) showed no effect of frequency, at least for frequencies above 400 Hz (Shailer and Moore 1987).

The duration threshold for the discrimination of time-reversed sinusoid pulses in the experiment of Green (1973) was approximately the same (at around 1 to 2 ms) for sinusoid frequencies of 1, 2, and 4 kHz, although the threshold for the 1-kHz condition was slightly higher than those for the other two.

In the case of the TMTF, the use of broadband noise carriers precludes direct statements about spectral effects in temporal processing and, indeed, raises the difficult question about what spectral region(s) is being used to detect the modulation. In an effort to address these issues, several investigators have bandpass filtered the noise after modulation and have studied the effects of center frequency and bandwidth (Rodenburg 1977; Viemeister 1977, 1979; van Zanten 1980). Although band limiting *after* modulation is necessary to eliminate spectral cues, this manipulation introduces its own set of problems. Such filtering will noticeably reduce the modulation depth of the input when the modulation frequency approaches half the bandwidth of the filter. Filtered noise also has a quasi-periodic envelope that may affect the detectability of the imposed sinusoidal modulation. Finally, in most studies, the effective bandwidth increased with center frequency, thus preventing clear separation of these potential effects. Eddins (1993) measured TMTFs using digitally generated SAM noise for which the nominal absolute bandwidth did not vary with center frequency. He found that the time constant associated with the TMTF did not vary with center frequency (600 to 4400 Hz) but that it decreased monotonically as the bandwidth increased from 200 to 1600 Hz. The effects of bandwidth are consistent with TMTF data from the earlier studies, but it is not clear how to account for these effects. They do not appear to result from reduction of the modulation depth of the input due to filtering.

Plack and Moore (1990) found an increase in the size of the temporal window at 300 Hz, measured using forward and backward masking, but concluded that the effect was actually too large to be accounted for by ringing in the filters. They argued instead that a more central component was largely responsible for the increase.

6.1.2 Temporal Resolution in Impaired Ears

Studies of temporal resolution in listeners with sensorineural hearing loss can be used to cast light on the role of peripheral filtering in temporal resolution. Impaired ears usually have broader than normal auditory filters (Glasberg and Moore 1986), and it follows from this that the temporal response of these filters should be shorter than normal. If peripheral filtering is important in determining the temporal resolution of the auditory system, then it might be expected that hearing-impaired listeners should actually show *superior* temporal resolution compared to normals.

Several studies have examined resolution in subjects with sensorineural hearing-impairment. In general, the results seem to depend on whether the comparison between impaired and normal ears is made at equal sound pressure level (SPL) or equal sensation level (SL). In the latter case, the level of the stimuli is adjusted to be the same number of decibels above absolute threshold in the normal and the impaired ears. The difference between these results largely reflects the fact that temporal resolution deteriorates at low SL (see Section 6.2).

The smallest detectable duration for gaps defined by noise markers has been shown to be longer for impaired ears than for normal ears when the comparison is made at equal SPL, although this result is not always found for comparisons at equal SL (Fitzgibbons and Wightman 1982; Tyler et al. 1982; Florentine and Buus 1984; Buus and Florentine 1985; Glasberg, Moore and Bacon 1987). However, Moore and Glasberg (1988) showed that, if sinusoids are used as markers the differences between normal and impaired ears disappear at equal SPL. Their data indicated a slight advantage for the impaired ears if the thresholds were measured at equal SL, which is consistent with the prediction based on the shorter temporal response of the impaired filters. This finding may, however, also reflect the fact that intensity discrimination is better for impaired than for normal ears at low SLs (Glasberg and Moore 1989), which is consistent with the hypothesis that gap detection involves an intensity discrimination component (see Section 5.).

Irwin and Purdy (1982) found that the detection of brief decrements in noise as a function of decrement depth was worse for impaired subjects than for normals, but the comparison was not made at equal SL and the noise level used in their experiment was close to absolute threshold for some of their subjects. Plack and Moore (1991) compared decrement detection for the impaired and normal ears of unilaterally hearing-impaired subjects, using noise levels in the normal ear at equal SL and SPL to the level in the impaired ear for each subject. They found that only one out of three impaired subjects showed an impairment in decrement detection in the impaired ear compared to the normal ear. On the other hand, they did not find any evidence that performance for the impaired ear was superior for the other two subjects.

TMTFs for impaired ears, measured using wideband noise carriers, have larger time constants than those for normals. This might be a consequence of the reduced listening bandwidth available to impaired ears because of high-frequency hearing loss (Bacon and Viemeister 1985). Simulating a reduced listening bandwidth in normal subjects by adding a high-pass noise to the modulated carrier also results in a larger time constant for the TMTF (Bacon and Viemeister 1985).

Another potentially interesting group of impaired subjects are individuals with cochlear implants. These listeners have essentially no frequency selectivity, the auditory filter being effectively bypassed by the stimulating electrode. Again, if the auditory filter is responsible for limiting resolution, performance might be expected to be superior for ears with implants. Gap thresholds for these subjects measured with sinusoidal markers are either similar to or higher than those for normals (Moore and Glasberg 1988). Shannon (1992) found that TMTFs measured in cochlear implant subjects had higher cutoff frequencies than those from normals, suggesting superior temporal resolution. The basis of the comparison may not be appropriate, however. It is very difficult to determine how sound intensities, specifically modulation depth, should be equated between the two groups of subjects.

It is clear that there is little evidence that impaired listeners have different temporal resolution than normals, although this does not necessarily imply that the response of the auditory filters is not an important limitation to temporal resolution. It could be that there are other components of hearing impairment that affect temporal resolution and act to counterbalance the resolution advantages of broad (or nonexistent) auditory filters.

In summary, there is no conclusive evidence that the auditory filters have a significant effect on temporal resolution in most conditions. Over very

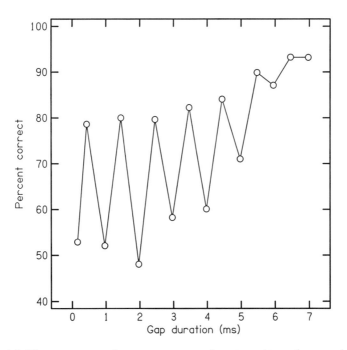

FIGURE 4.4. The percentage of correct responses in a two alternative gap detection task as a function of the duration of the gap. The stimulus was a 1-kHz sinusoid. The starting phase of the sinusoid after the gap was the same as the final phase before the gap. (Data are from Shailer and Moore 1987.)

short time intervals (less than 5 ms), however, the temporal effects of the filters are measurable. Shailer and Moore (1987) measured the detectability (in percent correct responses) of temporal gaps in sinusoids as a function of gap duration. They found that, in conditions where the sinusoid always started after the gap with the same phase it had when it was interrupted, the psychometric function showed oscillations at the frequency of the sinusoid. Figure 4.4 shows a typical example of the results they obtained. This effect can be explained on the basis of the ringing of the auditory filter: each time the gap duration was halfway between an integer number of periods of the sinusoid, the second marker interfered destructively with the ringing of the filter, producing a large dip in level and, hence, increased detectability. The oscillations observed by Shailer and Moore were largest at low frequencies, reflecting the longer temporal response of the auditory filter in this region. Moore et al. (1989) showed that these oscillations were markedly reduced in impaired ears, presumably reflecting the broad bandwidth (and, therefore, short temporal response) of the auditory filter in these cases. Furthermore, Moore and Glasberg (1988) did not observe these oscillations in the psychometric functions of cochlear implant patients, which is consistent with the idea that the effect is dependent on the auditory filter.

6.2 Temporal Resolution as a Function of Level

Gap detection studies have generally not shown a large effect of level, except at low SL. Plomp (1964) found poor performance at low sound levels with noise band markers, but there was little variability in performance for noise levels greater than 30 dB above absolute threshold. Moore and Glasberg (1988) obtained similar results using sinusoidal markers.

The form of the TMTF for broadband noise carriers appears to be independent of level over a wide range of levels (see Bacon and Viemeister 1985), suggesting that temporal resolution, as measured by the TMTF, is level independent. Modulation thresholds do increase, however, as the carrier level is reduced below a spectrum level of 0 dB SPL (about 35 dB SL). This increase in modulation threshold is somewhat surprising considering that for the detection of an increment in continuous noise, a situation closely related to modulation detection, Weber's Law holds down to levels of near 10 dB SL (Hanna, von Gierke, and Green 1986; Viemeister and Bacon 1988). For sinusoidal carriers and modulation frequencies for which sideband detection is not a problem, modulation thresholds decrease with increasing level, a result consistent with the "near miss" to Weber's Law for the intensity discrimination of pure tones. Although both intensity discrimination, especially increment detection, and modulation detection involve detection of amplitude changes, it is not clear how they are related. It is clear, however, that the frequently asserted relationship, dating back to Riesz's (1928) classic experiment, that the intensity JND is equal to peak-trough difference in power, is incorrect (see Bacon and Viemeister 1985).

Temporal window shapes measured using combined forward and backward masking are affected by stimulus intensity even at high levels, becoming narrower with increasing level (Moore et al. 1988; Plack and Moore 1990). These changes probably reflect nonlinearities in forward masking (see Section 8.1.1) that may be the result of neural adaptation.

It would appear, therefore, that temporal resolution is roughly independent of the stimulus level, at least in situations where forward masking does not play a major role. This level independence should not be taken to indicate, however, that the system is approximately envelope linear. Indeed, there is a sense in which the data indicate a strong nonlinearity, one not explicitly considered in models of temporal resolution. Specifically, the fact that the modulation threshold, expressed as a *relative* measure, is constant over a wide range of carrier levels means that proportionally larger amplitude fluctuations are required at higher carrier levels. This, of course, is analogous to Weber's Law for intensity discrimination. As noted above, intensity discrimination and temporal resolution seem to be closely related; temporal resolution requires intensity (or amplitude) discrimination. Until more is known about the underlying basis for Weber's Law, the typical modeling strategy of simply assuming Weber's Law by postulating, for example, that threshold corresponds to a constant max/min *ratio* seems acceptable, if only by default.

7. Across-Channel Temporal Resolution

In the research discussed so far, which has been concerned with within-channel resolution, it is generally assumed that differential stimulation across frequency is not a relevant dimension. That is, although there may be information available over many frequency channels, as in the case of resolving gaps in broadband noise or in the case of TMTFs measured with broadband noise carriers, it is assumed that differences in stimulation between channels are either nonexistent or are not used to perform the resolution task. Spectral cues, such as those produced by the modulation of a narrowband carrier would, of course, violate this assumption.

A perhaps more important issue concerns the ability to resolve temporal events occurring across frequency channels. Such across-channel resolution would seem to be important for detecting dynamic spectral changes such as those occurring with frequency sweeps or formant transitions. An example of across-channel resolution is the detection of differences in the onset time of tones of different frequencies. Pisoni (1977) studied the discrimination of the relative onset times between tones of 500 Hz and 1.5 kHz. He showed, for example, approximately 78% correct discrimination between simultaneous onset and an onset disparity of 20 ms. Unpublished work in our laboratory using more highly trained subjects and minimal uncertainty regarding stimulus conditions showed "onset disparity thresholds" (ODTs) of approximately 5 ms for tones of 1 and 4 kHz. The ODTs were the same regardless of which tone was delayed and appeared to depend only slightly on the relative spacing of the components.

Another approach for studying across-channel resolution uses Huffman sequences (Huffman 1962), a class of signals that, for a given length, have identical energy spectra but different phase spectra. Green (1973) used brief Huffman sequences, 12.8 ms or less, for which the energy at a certain frequency was delayed systematically. He found that additional delays of about 2 ms were discriminable from a standard delay that ranged from 1.6 to 6.4 ms. The delay thresholds were approximately constant when the frequency region containing the delayed energy was varied from 650 Hz to 4.2 kHz. Discrimination between different delays presumably involved an across-frequency comparison. It is not clear, however, whether within-channel cues also played a role in these experiments.

Formby and Forrest (1991) measured gap thresholds between sinusoidal "markers" that differed in frequency. When the markers differed substantially in frequency, these experiments could be interpreted as reflecting across-channel resolution, in this case offset-to-onset across-channel resolution. Formby and Forrest found, as did Williams and Perrott (1972) in a similar experiment, that the gap threshold increased as the frequency difference increased; thresholds of 100 ms were observed for frequency ratios of 1.5 at lower frequencies ranging from 500 Hz to 4 kHz. As the authors note, however, interpretation of these results requires a better understanding of the possible confounding effects of gating transients and of changes in overall duration.

Stimulated by the research on comodulation masking release (CMR; see Yost and Sheft, Chapter 6), there have been several experiments examining the detectability of phase disparities between the envelopes of simultaneously presented SAM tones with differing carrier frequencies (for a review see Green, Richards, and Onsan 1990). For widely spaced carrier frequencies, detection of such disparities could be determined by the ability to resolve temporal differences across frequency and, indeed, this is the usual interpretation. Although there are substantial differences between the techniques used in these studies, there is general agreement that, for modulation frequencies below 100 to 200 Hz and large modulation depths, fairly small phase disparities, approximately 20 to 100 degrees, are detectable. For example, Strickland et al. (1989) found that, with 100%, 128-Hz SAM tones at carrier frequencies of 1 and 3.2 kHz, an envelope phase difference of approximately 75 degrees was just discriminable from inphase envelopes when the carrier levels were both 65 dB SPL. This corresponds to a time difference between envelope peaks of 1.6 ms. In general, phase disparity thresholds or "coherence thresholds" are not markedly dependent on the center frequency or on the spacing of the carriers once the spacing exceeds about an octave.

An ever-present issue when concerned with across-channel resolution is the possible role of within-channel cues. In most across-channel tasks, a potentially useful strategy for the observer is to use a single frequency channel that is placed so as to maximize envelope interactions. In the experiment of Strickland et al. (1989), for example, the observer might perform the task by detecting a change in the envelope in a single channel whose center frequency is between 1 and 3.2 Hz. There is considerable evidence, however, that a simple within-channel model does not provide a general explanation for the results of these experiments. Goldstein (1966), for example, showed that envelope phase disparity thresholds were largely unaffected when a bandpass noise was presented between the two carrier frequencies. Similarly, the independence of the resolution measures on frequency spacing argues against a simple within-channel explanation. Nevertheless, some caution seems appropriate in inferring that across-channel, not within-channel, mechanisms are involved. The problem is that the within-channel explanations may be too simple. Specifically, such explanations typically consider only the effects of a single, linear channel that allows simple, additive waveform interactions to occur. There are, however, unexplained "within-channel" interactions that are inconsistent with such a view. Recall that Wakefield and Viemeister (1985) showed strong, nonadditive envelope interactions over a very large frequency range. The phenomenon of MDI could also be interpreted as reflecting within-channel interactions over a large frequency range. To offer a rather extreme view, it may be that all across-channel resolution is mediated by within-channel mechanisms and that no true across-channel comparison actually occurs.

8. Temporal Aspects of Masking

8.1 Nonsimultaneous Masking

Nonsimultaneous masking refers to the decrease in the detectability of a signal caused by a masking stimulus which is not physically present at the time of presentation of the signal. Forward masking refers to the deterioration in performance caused by a masker preceding the signal, and backward masking to the case where the signal precedes the masker.

8.1.1 Forward Masking and Adaptation

The detection threshold for a signal is elevated when the signal is presented shortly after a masking stimulus of a similar frequency. The shorter the delay between masker and signal, the greater the threshold elevation. Forward masking might be the result of the ear's limited temporal resolution. Temporal smearing (such as that produced by a leaky integrator) would increase the effective duration of the masker, and so the signal would be masked "directly" at some later stage in the auditory pathway (Smiarowski and Carhart 1975). However, several researchers have proposed that the principle cause of forward masking is neural adaptation (Duifhuis 1973; Smith 1977; Kidd and Feth 1982), except perhaps for short masker-signal delays where the amount of forward (and backward) masking may be determined by the overlap of responses on the basilar membrane (Duifhuis 1973). In response to a sound with a sudden onset, a single auditory neuron shows an initially high rate of firing, followed by a decline in activity (Kiang et al. 1965). If the sound is terminated, the normal spontaneous activity of the neuron in quiet is suppressed, showing a gradual recovery over a period of 150 ms or so. This process might result in psychophysical forward masking by reducing the neural response to a signal presented after a masker.

This hypothesis is supported by evidence that some of the characteristics of forward masking seem to parallel the physiological data on adaptation (see also Yost and Shoft, Chapter 6). Figure 4.5 illustrates the rate of recovery from forward masking as a function of the intensity of the masking stimulus, in this case a 1-kHz sinusoid. Two of the main features of forward masking are apparent. First, the rate of recovery from forward masking increases with masker level. If the curves are extrapolated to the right, they converge at a signal delay of about 200 ms. This is similar to the time taken for neurons to recover from adaptation (Smith, Brachman and Goodman 1983). Related to this is the finding that, unlike in simultaneous masking where Weber's Law

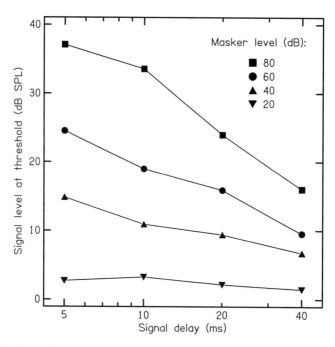

FIGURE 4.5. Detection thresholds for a 20-ms, 1-kHz sinusoidal signal presented after a 300-ms, 1-kHz sinusoidal masker (10-ms onset and offset ramps in each case). Signal delay refers to the time between the offset of the masker and the onset of the signal. The parameter is masker level. (Data are from Jesteadt, Bacon, and Lehman 1982.)

seems to apply, the threshold does not increase in proportion to the masker level. The four masker levels shown in the figure cover a range of 60 dB, yet the thresholds cover a range of less then 40 dB. Adaptation seems to show a similar compressive dependence on adaptor level (Smith 1979). An additional effect that is not illustrated in Figure 4.5 is that the amount of forward masking seems to increase with masker duration for durations up to 200 ms (Kidd and Feth 1982). This seems inconsistent with a short-time constant temporal process (although see Section 8.1.2) but could be explained by a relatively long-term adaptation.

Although these results suggest that there is a qualitative difference between forward and simultaneous masking that may be the result of adaptive processes, there are some difficulties with the adaptation account. Signal detectability is usually assumed to be based on both the neural activity due to the signal and competing neural noise. Adaptation reduces the spontaneous activity of the neuron as well as the response to the signal and, therefore, may not necessarily impair performance. Supporting this idea, Relkin and Turner (1988) compared the neural response to a signal presented after a tone with the neural response in the absence of the signal. They showed that the effect of peripheral adaptation in the chinchilla was not great enough to account for the behavioral forward masking thresholds.

One piece of psychophysical evidence suggests that forward masking cannot be entirely the result of peripheral adaptation. This is the finding that the binaural masking level difference (see Yost and Shaft, Chapter 6) can be observed in forward masking conditions (Small et al. 1972; Yost and Walton 1977). This would not be possible if the response to the signal had been removed by adaptation prior to binaural integration.

8.1.2 The Nonadditivity of Forward and Backward Masking

Although the effects of backward masking on suprathreshold tasks such as frequency discrimination can be considerable (see Massaro 1975), it is not clear that a backward masker in isolation has any substantial effect on signal detectability. In situations where there is a forward and backward masker, however, the effects of the backward masker are more significant. Bilger (1959) measured the detection of sinusoids under conditions of forward, backward, and combined forward and backward masking. He found that the amount of masking obtained in the combined case was greater than that predicted by summing the contributions of the forward and backward masking measured separately. This phenomenon is often referred to as the "nonadditivity" of forward and backward masking. Robinson and Pollack (1973) have given a qualitative explanation of the phenomenon based on the idea of the "temporal integrating period," which is equivalent to the temporal window concept described in Section 5.2. Briefly, it is suggested that the signal-to-masker ratio in the case of forward masking alone is improved by "listening" through a temporal window centered at a time after the signal presentation, which therefore integrates less masker energy than a temporal window centered on the signal. The explanation depends on having a window shape with a flat top and steep skirts; for an exponential window, the signal-to-masker ratio does not change. The addition of a backward masker reduces the effectiveness of this "off-time listening" because the reduction in the masking energy caused by moving the window away from the forward masker is accompanied by an increase in the energy integrated from the backward masker. Therefore, the signal threshold appears to be unusually high in the combined forward and backward masking case.

Although this hypothesis is appealing in some ways, it is not clear that off-time listening can quantitatively account for the experimental results. Another possibility is that the nonadditivity effect results from nonlinearities in the integration of intensity. Penner and Shiffrin (1980) proposed a model similar to that described in Section 5, in which the nonlinearity before the leaky integrator is highly compressive. This scheme seems to be able to account for the nonadditivity effect (Penner 1980), as well as the effect of masker duration on forward masking. It is worthwhile examining how this might work. Consider the case of a single instantaneous impulse. In the model of Penner and Shiffrin, doubling the intensity of the impulse will result in a less than two-fold increase in the output of the temporal window because of the compressive nonlinearity. Now, if instead of doubling the intensity at a single instant, another impulse of the same intensity is added, but at a different time, then the two impulses will be compressed independently prior to integration. Assuming the temporal window gives equal weight to each, in this case the output of the window will be double what it was for a single click. This means that, in some circumstances, more masking can be obtained by distributing the masker energy over time rather than concentrating it at a single instant. Therefore, combined forward and backward masking is relatively more effective than forward or backward masking in isolation.

8.2 Overshoot

The threshold for detecting a signal in the presence of a simultaneous noise masker is elevated when the signal is presented shortly after the onset of the masker, compared to a presentation in the temporal center of the noise. The elevation can be as much as 10 dB (Zwicker 1965a). This phenomenon has been termed "overshoot" or the "temporal effect." The amount of overshoot is largest at medium sound levels (Bacon 1990) and at high frequencies (Carlyon and White 1992). Overshoot also tends to be largest for brief signals (Elliott 1967). A related finding is that the threshold for brief signals in maskers of the same duration (gated) is higher than the threshold for signals presented in continuous maskers.

It has been suggested that overshoot is related to short-term neural adaptation (Green 1969; Bacon and Viemeister 1985). When a neuron is stimulated by a sound, it shows an initial high rate of firing that drops rapidly over the first 10 to 20 ms (Kiang et al. 1965). This might be related to the masking difference. Smith and Zwislocki (1971, 1975) have demonstrated that the firing rate of cochlear nucleus and auditory nerve units show a smaller *proportional* increase in response to a stimulus intensity increment at the onset of the stimulus than for times after. In support of this hypothesis is the finding that the threshold for a signal presented at the onset of a noise can be reduced by presenting a noise burst that ends shortly before the masker onset (Zwicker 1965a; Elliott 1969). This first noise might have the effect of "pre-adapting" the system.

Some characteristics of the overshoot phenomenon are inconsistent with an explanation in terms of simple adaptation, however. For instance, overshoot seems to be largest if the masker bandwidth is greater than one critical band (Zwicker 1965b). Furthermore, the threshold at masker onset can be reduced even if the spectral energy of a preceding noise burst is remote from the signal frequency, as in the case of a notched noise (Carlyon 1987; McFadden 1989). It could be that two different mechanisms are involved in the reduction of threshold (see Carlyon 1987) and, by inference, in the production of the threshold difference termed overshoot: an on-frequency effect that may be the result of simple adaptation in the auditory nerve, and an off-frequency effect termed "enhancement" wherein the notched noise emphasizes the frequency region of the notch and, hence, improves the detectability of the signal, thereby reducing overshoot. Enhancement has been measured as a shift in the perceived spectral balance of a signal towards the complementary spectrum of a preceding adaptor (Viemeister 1980; Summerfield et al. 1984). Another possibility is that the off-frequency effect reflects the pooling of neural responses across frequency, effectively increasing the width of the critical band. Champlin and McFadden (1989) suggested that there may be a detector in the auditory system that integrates synchronous activity across fibers with different characteristic frequencies. For wideband noise maskers, such a synchrony would occur at masker onset. The integration process would therefore cause an increase in the effective masker level at onset and a corresponding increase in the threshold for signals presented at this time.

9. Temporal Integration

9.1 Detection and Discrimination

While temporal resolution refers to the auditory system's ability to "hear out" events happening over a very short time, temporal integration or temporal summation refers to the ability to combine information across a relatively long period of time in order to improve performance. It is generally true in audition, as in other senses, that detection and discrimination performance improves with increases in duration for durations up to several hundred milliseconds.

Most of the research on temporal integration has focused on detection tasks. For the detection of pure tones in quiet and in noise, the data relating signal threshold to duration are reasonably well described by the function proposed by Plomp and Bouman (1959): $I/I_\infty = (1 - e^{-T/v})^{-1}$, where I is the threshold in (linear) intensity units, T is the signal duration (sometimes corrected for rise-decay times), I_∞ is the threshold for long-duration signals, and v is the time constant used to fit the data. According to this function, for $T \ll v$, $I \propto 1/T$, i.e., signal *energy*, the time integral of intensity, is constant at threshold.[2] For signal durations less than 10 to 20 ms, it is usually observed that the energy of the signal must be increased with decreasing duration. This is thought to result from the "splatter" of signal energy to frequencies outside the critical band centered on the signal frequency (Garner 1947; Sheeley and Bilger 1964). Although there is some disagreement in the literature, it is generally found that the time constant decreases at higher signal frequencies and ranges from 300 to 500 ms for frequencies below 1 kHz to approximately 100 to 200 ms for frequencies above 5 kHz (for a recent review see Gerken, Bhat, and Hutchison-Clutter 1990).

As with detection, performance in intensity discrimination and frequency discrimination improves with increases in duration. In intensity discrimination, however, the rate of decrease of the JND is considerably less than for detection. For gated pure tones (see Florentine 1986) and for gated broadband noise (see Viemeister 1988), the JND decreases by 3 to 4 dB per decade increase in duration, in contrast to the approximately 10 dB/decade seen for detection. Furthermore, there is a substantial difference, at least for broadband noise, in the temporal integration functions for gated and continuous standards; the JND for increments in continuous noise decreases by about 7 dB/decade. The JNDs for continuous standards are 5 dB larger at 0.8 ms and about 3 dB smaller at 200 ms than those for gated standards. The poorer JNDs at short durations and, to some extent, the change in slope of the integration function may reflect "overintegration" with continuous standards and brief increments. This argument is, of course, similar to that underlying the "critical masking interval." In this regard, it is surprising that for pure tone detection in noise, the data do not point to a clear difference in the integration functions for gated and continuous maskers (see Wier et al. 1977).

There are other integration-like phenomena in audition. The frequency discrimination of pure tones improves with duration (Moore 1973; Goldstein and Srulovic 1977), loudness appears to increase with duration, although the temporal characteristics of this increase are uncertain (see Scharf 1978), and, in general, there is an improvement in performance in most auditory tasks with increases in duration, at least over some range of durations. Whether

---

2. It has been argued that the function $I = kT^{-\alpha}$, k a constant, better describes the integration data (Green 1985; Gerken, Bhat, and Hutchison-Clutter 1990). The values of α that fit the data are typically close to one; thus, signal energy still is approximately constant at threshold.

these integration phenomena are related is unclear. For the present purposes, it will be more useful to consider theoretical approaches to temporal integration only for detection. These may indeed provide a basis for a general theory of integration.

9.2 Theoretical Aspects of Temporal Integration

9.2.1 Classical Treatments

Most models of detection employ some form of integration or summation that occurs over an interval of several hundred milliseconds. Perhaps the clearest and most fully developed example is the energy detector (see Green and Swets 1966) in which it is assumed that the power of the (filtered) input is perfectly integrated over some time interval. If the integration period is fixed at, say, 200 ms, then the model predicts that, for detection of a tone in continuous noise, signal energy at threshold will be constant for durations of less than 200 ms and signal power will be constant for longer durations. This is in reasonably good agreement with the data (Green, Birdsall, and Tanner 1957). If the noise is gated or if the integration period matches the signal duration, the model predicts a square root dependence: $I\sqrt{T} = k$. This is inconsistent with the data for tones in noise but is at least roughly consistent with the data described in Section 9.1 on the intensity discrimination of noise.

The energy detector, essentially, accounts for the phenomenon of temporal integration by postulating that long-term integration actually occurs in auditory processing. There are other models similar in this regard except that they incorporate leaky integration or running averaging (Plomp and Bouman 1959; Zwislocki 1960, 1969; Jeffress 1967, 1968). These models differ considerably in their details but have in common long time-constant integration.

9.2.2 The Integration/Resolution Paradox

The leaky integration models used to describe temporal integration are similar to those used to describe temporal resolution. The major difference is in the time constants of integration: the 200 to 300 ms time constants suggested by the data on temporal integration contrast sharply with the short time constants, 3 to 10 ms, indicated by the data on temporal resolution. This difference is, essentially, the "temporal resolution-temporal integration paradox" (de Boer 1985; Green 1985). Figure 4.6 illustrates the problem. The left part of the figure shows that the response of a leaky integrator with a 3-ms time constant follows, or resolves, a 3-ms gap much better than the integrator with the 300-ms time constant. The center and right part of the figure shows that the 300-ms time constant integrator exhibits almost perfect temporal integration: the integrator output during the pulse is nearly proportional to the integral of the input, and the output at the end of the two equal energy pulses is approximately the same. The 3-ms integrator, however, shows no time-intensity trade; the intensity of the longer duration pulse would have

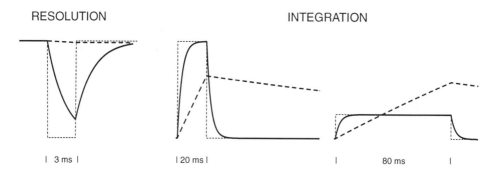

FIGURE 4.6. The resolution/integration paradox. The thin dashed lines are the (intensity) envelopes of the input. The thick dashed lines are the responses of a leaky integrator with a 300-ms time constant; the thick solid lines are for a 3-ms time constant. The figure for resolution represents detection of a gap in an otherwise continuous stimulus. The figures for integration represent detection of a 20-ms and an 80-ms stimulus with equal energy.

to be the same as that of the shorter pulse to maintain a constant output "threshold" at the end of the pulses.

The common approach to the resolution/integration paradox is to assume that the auditory system has two (or more) time constants and that the effective time constant depends on the task. A more appealing solution was proposed by Penner (1978) who argued that a compressive nonlinearity preceding short time-constant integration could yield a temporal integration function consistent with the data (see also Section 8.1.2).

9.2.3 The Multiple Look Model

A general issue is whether long time-constant integration, such as that proposed by the classical models, occurs in auditory processing. Viemeister and Wakefield (1991) presented evidence that it does not, at least not in detection tasks. They showed that the threshold for a pair of brief pulses indicated classic power integration only for pulse separations of less than 5 ms; for longer separations, the thresholds did not change with separation and the pulses appeared to be processed independently (cf. Zwislocki, Hellman, and Verrillo 1962). In another experiment, Viemeister and Wakefield (1991) showed that the threshold for a pair of tone pulses was lower than for a single pulse, indicating some type of integration, but was not affected by changes in the level of the noise that was presented between the pulses. These data are inconsistent with long-term integration. The results of the second experiment are also inconsistent with the model proposed by Penner (1978), discussed in Sections 9.2.2 and 8.1.2.

To account for these data and, more generally, the basic phenomenon of temporal integration, Viemeister and Wakefield (1991) proposed a "multiple look" model: "looks" or samples from a short time-constant process are stored in memory and can be accessed and processed selectively, depending on the task. The selective processing and short time constant allows the model to account for the data from the pulse pair experiments described above and, in principle, for the data on temporal resolution. In a detection task, the improvement in performance with increasing duration simply reflects an increase in the number of looks. Temporal integration, then, may result from the combination of information from different looks, not from true long-term integration. The predicted temporal integration function depends upon how the looks are weighted and combined. Viemeister and Wakefield (1991) demonstrated that, with appropriate weighting and combination, the model can yield predictions consistent with the integration data.

The multiple look model is essentially an extension of the information integration model used to describe the detection of multicomponent signals and of multiply presented signals (see Green and Swets 1966). It appears to offer a solution to the temporal resolution-temporal integration paradox and provides a framework for describing a wide range of temporal phenomena including, of course, temporal integration. There are, however, many uncertainties about the model, including those issues discussed in the context of temporal resolution. In accounting for temporal integration, a clear weakness of the model is the ad hoc nature of the weighting function. While some sort of weighting and combination seems reasonable, it certainly would be desirable to have an independent experimental measurement of how information is actually weighted and combined in a detection task with long-duration signals.

10. Summary

It is evident, even from this limited review, that our knowledge about certain aspects of temporal processing is quite extensive. The recent emphasis of research in this general area has been on temporal resolution, especially on within-channel resolution. There is fairly good agreement over a large number of studies on the basic properties of within-channel resolution. The time constant that usually arises from both simple acuity measures such as gap detection and time constants estimated from TMTFs or temporal windows is approximately 3 to 8 ms. Similarly, there is general agreement that within-channel resolution is not markedly dependent on frequency region or on level. The simple leaky integrator/envelope detector appears to provide an adequate, first-approximation model of within-channel temporal processing. A major problem is the indication, both from data and theory, that resolution is not limited by auditory filtering. It is not clear how the auditory system can overcome what would appear to be such a fundamental limitation in some circumstances.

The leaky integration model, with a time constant on the order of a few milliseconds, may provide the basis for a general account of temporal processing, especially when coupled with more sophisticated "central" processing. Such a scheme was invoked, for example, to account for the phenomenon of temporal integration and thereby offered a solution to the resolution/integration paradox. It would also appear to be useful to extend this basic model to account for across-channel temporal processing. It is not clear how across-channel resolution is dependent on within-channel processing, although it seems likely that such a dependency exists. The time constant for across-channel resolution is more elusive than that for within-channel resolution but seems to have a similar value. One interpretation is that the data do not truly reflect across-channel resolution, perhaps because of interactions that have been shown to occur between widely spaced frequency regions. Similarly, it may be that the within- versus across-channel distinction is invalid and that all temporal resolution reflects within-channel processes.

Although elaboration and refinement of the simple leaky integration model seems appropriate, there are important temporal phenomena that appear to reflect processes outside of the domain of the model. These include overshoot, MDI, forward masking, and certain "enhancement" effects. It is possible that a general account of these phenomena, perhaps incorporating adaptation, can be developed and reconciled with the simple leaky integration scheme. It is clear, however, that we are presently far from having anything approaching a unified account even of the presumably basic, "low level," phenomena discussed in this chapter.

Acknowledgments. We thank Christine Mason and Elizabeth Strickland for comments on this manuscript. We also thank Dr. Constantine Trahiotis for stimulating discussions on the real meaning of "envelope." This work was partially supported by NIDCD Grant No. DC00683.

References

Bacon SP (1990) Effect of masker level on overshoot. J Acoust Soc Am 88:698–702.

Bacon SP, Grantham DW (1989) Modulation masking: Effects of modulation frequency, depth, and phase. J Acoust Soc Am 85:2575–2580.

Bacon SP, Viemeister NF (1985) Temporal modulation transfer functions in normal-hearing and hearing-impaired listeners. Audiology 24:117–134.

Bilger RC (1959) Additivity of different types of masking. J Acoust Soc Am 31:1107–1109.

Burns EM, Viemeister NF (1981) Played-again SAM: Further observations on the pitch of amplitude-modulated noise. J Acoust Soc Am 70:1655–1660.

Buunen TJF, van Valkenburg DA (1979) Auditory detection of a single gap in noise. J Acoust Soc Am 65:534–537.

Buus S, Florentine M (1985) Gap detection in normal and impaired listeners: The effect of level and frequency. In: Michelson A (ed) Time Resolution in Auditory Systems. Berlin, Heidelberg: Springer-Verlag, pp. 159–179.

Carlyon RP (1987) A release from masking by continuous, random, notched noise. J Acoust Soc Am 81:418–426.

Carlyon RP, White LJ (1992) Effect of signal frequency and masker level on the frequency regions responsible for the overshoot effect. J Acoust Soc Am 91:1034–1041.

Champlin CA, McFadden D (1989) Reductions in overshoot following intense sound exposures J Acoust Soc Am 85:2005–2011.

de Boer E (1985) Auditory time constants: A paradox? In: Michelson A (ed) Time Resolution in Auditory Systems. Berlin, Heidelberg: Springer-Verlag, pp. 141–158.

Duifhuis H (1973) Consequences of peripheral frequency selectivity for nonsimultaneous masking. J Acoust Soc Am 54:1471–1488.

Eddins DA (1993) Amplitude modulation detection of narrowband noise: Effects of absolute bandwidth and frequency region. J Acoust Soc Am, 93:470–479.

Elliott LL (1967) Development of auditory narrowband frequency contours. J Acoust Soc Am 42:143–153.

Elliott LL (1969) Masking of tones before, during, and after brief silent periods in noise. J. Acoust Soc Am 45:1277–1279.

Festen JM (1987) Speech-reception threshold in a fluctuating background sound and its possible relation to temporal auditory resolution. In: Schouten MEH (ed) The Psychophysics of Speech Perception. Dordrecht, The Netherlands: Nijhoff, pp. 461–466.

Festen JM, Plomp R (1981) Relations between auditory functions in normal hearing. J Acoust Soc Am 70:356–369.

Festen JM, Plomp R (1983) Relations between auditory functions in impaired hearing. J Acoust Soc Am 73:652–662.

Festen JM, Houtgast T, Plomp R, Smoorenburg GF (1977) Relations between interindividual differences of auditory functions. In: Evan EF, Wilson JP (eds) Psychophysics and Physiology of Hearing. London: Academic Press, pp. 311–319.

de Filippo CL, Snell KB (1986) Detection of a temporal gap in low-frequency narrowband signals by normally hearing and hearing-impaired subjects. J Acoust Soc Am 80:1354–1358.

Fitzgibbons PJ (1983) Temporal gap detection in noise as a function of frequency, bandwidth, and level. J Acoust Soc Am 74:373–379.

Fitzgibbons PJ, Wightman FL (1982) Gap detection in normal and hearing-impaired listeners. J Acoust Soc Am 72:761–765.

Fletcher H (1940) Auditory patterns. Rev Mod Phys 12:47–65.

Florentine M (1986) Level discrimination of tones as a function of duration. J Acoust Soc Am 79:792–798.

Florentine M, Buus S (1984) Temporal gap detection in sensorineural and simulated hearing impairment. J Speech Hear Res 27:449–455.

Formby C, Forrest TG (1991) Detection of silent temporal gaps in sinusoidal markers. J Acoust Soc Am 89:830–837.

Forrest TG, Green DM (1987) Detection of partially filled gaps in noise and the temporal modulation transfer function. J Acoust Soc Am 82:1933–1943.

Garner WR (1947) The effect of frequency spectrum on temporal integration of energy in the ear. J Acoust Soc Am 19:808–815.

Gerken GM, Bhat VKH, Hutchison-Clutter MH (1990) Auditory temporal integration and the power-function model. J Acoust Soc Am 88:767–778.

Glasberg BR, Moore BCJ (1986) Auditory filter shapes in subjects with unilateral and bilateral cochlear impairments. J Acoust Soc Am 79:1020–1033.

Glasberg BR, Moore BCJ (1989) Psychoacoustical abilities of subjects with unilateral and bilateral cochlear hearing impairments and their relationship to the ability to understand speech. Scand Audiol Suppl 32:1–25.

Glasberg BR, Moore BCJ, Bacon SP (1987) Gap detection and masking in hearing-impaired and normally hearing subjects. J Acoust Soc Am 81:1546–1556.

Goldstein JL (1966) An investigation of monaural phase perception. Doctoral Dissertation, The University of Rochester, Rochester, NY. Ann Arbor, MI: University Microfilms, Publ. No. 66-6852.

Goldstein JL, Srulovic P (1977) Auditory nerve spike intervals as an adequate basis for aural frequency measurement. In: Evans EF, Wilson JP (eds) Psychophysics and Physiology of Hearing. London; Academic Press, pp. 337–346.

Green DM (1969) Masking with continuous and pulsed sinusoids. J Acoust Soc Am 46:939–946.

Green DM (1973) Temporal acuity as a function of frequency. J Acoust Soc Am 54:373–379.

Green DM (1985) Temporal factors in psychoacoustics In: Michelson A (ed) Time Resolution in Auditory Systems. Berlin, Heidelberg: Springer-Verlag, pp. 122–140.

Green DM, Swets JA (1966) Signal Detection Theory and Psychophysics. New York: John Wiley and Sons.

Green DM, Birdsall TG, Tanner WP Jr (1957) Signal detection as a function of signal intensity and duration. J Acoust Soc Am 29:523–531.

Green DM, Richards VM, Onsan ZA (1990) Sensitivity to envelope coherence. J Acoust Soc Am 87:323–329.

Green GG, Kay, RH (1973) The adequate stimuli for channels in the human auditory pathways concerned with the modulation present in frequency-modulated tones. J Physiol (London) 234:50–52.

Green GG, Kay RH (1974) Channels in the human auditory system concerned with the waveform of modulation present in amplitude- and frequency-modulated tones. J Physiol (London) 241:29–30.

Hanna TE, von Gierke SM, Green DM (1986) Detection and intensity discrimination of a sinusoid. J Acoust Soc Am 80:1335–1340.

Henning GB, Gaskell H (1981) Monaural phase sensitivity measured with Ronken's paradigm. J Acoust Soc Am 70:1669–1673.

Houtgast T (1989) Frequency selectivity in amplitude modulation detection. J Acoust Soc Am 85:1676–1680.

Huffman DA (1962) The generation of impulse-equivalent pulse trains. IRE Transactions IT 8:S10–S16.

Irwin RJ, Purdy SC (1982) The minimum detectable duration of auditory signals for normal and hearing-impaired listeners. J Acoust Soc Am 71:967–974.

Jeffress LA (1967) Stimulus-oriented approach to detection re-examined. J Acoust Soc Am 41:480–488.

Jeffress LA (1968) Mathematical and electrical models of auditory detection. J Acoust Soc Am 44:187–203.

Jesteadt W, Bacon SP, Lehman JR (1982) Forward masking as a function of frequency, masker level, and signal delay. J Acoust Soc Am 71:950–962.

Kiang NY-S, Watanabe T, Thomas EC, Clark LF (1965) Discharge Patterns of Single Fibers in the Cat's Auditory Nerve. Research Monograph No. 35. Cambridge, MA: MIT Press.

Kidd G, Jr, Feth LL (1982) Effects of masker duration in pure-tone forward masking. J Acoust Soc Am 72:1384–1386.

Leshowitz B (1971) The measurement of the two-click threshold. J Acoust Soc Am 49:462–466.
Massaro DW (1975) Backward recognition masking. J Acoust Soc Am 58:1059–1065.
McFadden D (1989) Spectral differences in the ability of temporal gaps in noise to reset the mechanisms underlying overshoot. J Acoust Soc Am 85:254–261.
Moody DB, Cole D, Davidson LM, Stebbins WC (1984) Evidence for a reappraisal of the psychophysical selective adaptation paradigm. J Acoust Soc Am 76:1076–1079.
Moore BCJ (1973) Frequency difference limens for short-duration tones. J Acoust Soc Am 54:610–619.
Moore BCJ, Glasberg BR (1988) Gap detection with sinusoids and noise in normal, impaired, and electrically stimulated ears. J Acoust Soc Am 83:1093–1101.
Moore BCJ, Glasberg BR, Plack CJ, Biswas AK (1988) The shape of the ear's temporal window. J Acoust Soc Am 83:1102–1116.
Moore BCJ, Glasberg BR, Donaldson E, McPherson T, Plack CJ (1989) Detection of temporal gaps in sinusoids by normally hearing and hearing-impaired subjects. J Acoust Soc Am 85:1266–1275.
Panter PF (1965) Modulation, Noise and Spectral Analysis. New York: McGraw-Hill.
Patterson RD (1976) Auditory filter shapes derived with noise stimuli. J Acoust Soc Am 59:640–654.
Penner MJ (1977) Detection of temporal gaps in noise as a measure of the decay of auditory sensation. J Acoust Soc Am 61:552–557.
Penner MJ (1978) A power law transformation resulting in a class of short-term integrators that produce time-intensity trades for noise bursts. J Acoust Soc Am 63:195–201.
Penner MJ (1980) The coding of intensity and the interaction of forward and backward masking. J Acoust Soc Am 67:608–616.
Penner MJ, Cudahy E (1973) Critical masking interval: A temporal analog of the critical band. J Acoust Soc Am 54:1530–1534.
Penner MJ, Shiffrin RM (1980) Nonlinearities in the coding of intensity within the context of a temporal summation model. J Acoust Soc Am 67:617–627.
Penner MJ, Robinson CE, Green DM (1972) The critical masking interval. J Acoust Soc Am 52:1661–1668.
Pierce JR, Lipes R, Cheetham C (1977) Uncertainty concerning the direct use of time information in hearing: Place clues in white-spectra stimuli. J Acoust Soc Am 61:1609–1621.
Pisoni DB (1977) Identification and discrimination of the relative onset time of two component tones: Implications for voicing perception in stops. J Acoust Soc Am 61:1352–1361.
Plack CJ, Moore BCJ (1990) Temporal window shape as a function of frequency and level. J Acoust Soc Am 87:2178–2187.
Plack CJ, Moore BCJ (1991) Decrement detection in normal and impaired ears. J Acoust Soc Am 90:3069–3076.
Plomp R (1964) The rate of decay of auditory sensation. J Acoust Soc Am 36:277–282.
Plomp R, Bouman MA (1959) Relation between hearing threshold and duration for tone pulses. J Acoust Soc Am 31:749–758.

Rees A, Møller A (1983) Responses of neurons in the inferior colliculus of the rat to AM and FM tones. Hear Res 10:301–330.

Relkin EM, Turner CW (1988) A reexamination of forward masking in the auditory nerve. J Acoust Soc Am 84:584–591.

Riesz RR (1928) Differential intensity sensitivity of the ear for pure tones. Phys Rev 31:867–875.

Rice SO (1982) Envelopes of narrow band signals. Proc IEEE 70:692–699.

Robinson CE, Pollack I (1973) Interaction between forward and backward masking: A measure of the integrating period of the auditory system. J Acoust Soc Am 53:1313–1316.

Rodenburg M (1977) Investigation of temporal effects with amplitude modulated signals. In: Evans EF, Wilson JP (eds) Psychophysics and Physiology of Hearing. London: Academic Press, pp. 429–437.

Ronken D (1970) Monaural detection of a phase difference between clicks. J Acoust Soc Am 47:1091–1099.

Scharf B (1978) Loudness. In: Carterette EC, Friedman MP (eds) Handbook of Perception, Volume IV. New York: Academic Press, pp. 187–242.

Schreiner CE, Urbas JV (1988) Representation of amplitude modulation in the auditory cortex of the cat. II. Comparison between cortical fields. Hear Res 32:49–65.

Shailer MJ, Moore BCJ (1983) Gap detection as a function of frequency, bandwidth, and level. J Acoust Soc Am 74:467–473.

Shailer MJ, Moore BCJ (1985) Detection of temporal gaps in bandlimited noise: Effects of variations in bandwidth and signal-to-masker ratio. J Acoust Soc Am 77:635–639.

Shailer MJ, Moore BCJ (1987) Gap detection and the auditory filter: Phase effects using sinusoidal stimuli. J Acoust Soc Am 81:1110–1117.

Shannon RV (1992) Temporal modulation transfer functions in patients with cochlear implants. J Acoust Soc Am 91:2156–2164.

Sheeley EC, Bilger RC (1964) Temporal integration as a function of frequency. J Acoust Soc Am 36:1850–1857.

Sheft S, Yost WA (1990) Temporal integration in amplitude modulation detection. J Acoust Soc Am 88:796–805.

Small AM, Boggess J, Klich R, Kuehn D, Thelin J, Wiley T (1972) MLD's in forward and backward masking. J Acoust Soc Am 51:1365–1367.

Smiarowski RA, Carhart R (1975) Relations between temporal resolution, forward masking, and simultaneous masking. J Acoust Soc Am 57:1169–1174.

Smith RL (1977) Short-term adaptation in auditory nerve fibers: Some poststimulatory effects. J Neurophysiol 40:1098–1112.

Smith RL (1979) Adaptation, saturation, and physiological masking in single auditory-nerve fibers. J Acoust Soc Am 65:166–178.

Smith RL, Zwislocki JJ (1971) Responses of some neurons in the cochlear nucleus to tone-intensity increments. J Acoust Soc Am 50:1520–1525.

Smith RL, Zwislocki JJ (1975) Short-term adaptation and incremental responses in single auditory-nerve fibers. Biol Cybernet 17:169–182.

Smith RL, Brachman ML, Goodman DA (1983) Adaptation in the auditory periphery. Ann NY Acad Sci 405:79–93.

Strickland EA, Viemeister NF, Fantini DA, Garrison MA (1989) Within- versus cross-channel mechanisms in detection of envelope phase disparity. J Acoust Soc Am 86:2160–2166.

Summerfield Q, Haggard M, Foster J, Gray S (1984) Perceiving vowels from uniform spectra: Phonetic exploration of an auditory aftereffect. Percept Psychophys 35: 203–213.
Tansley BW, Suffield JB (1983) Time course of adaptation and recovery of channels selectively sensitive to frequency and amplitude modulation. J Acoust Soc. Am 74:765–775.
Tyler RS, Summerfield Q, Wood EJ, Fernandes MA (1982) Psychoacoustic and phonetic temporal processing in normal and hearing impaired listeners. J Acoust Soc Am 72:740–752.
van Zanten GA (1980) Temporal modulation transfer functions for intensity modulated noise. In: van den Brink G, Bilsen FA (eds) Psychophysical and Behavioral Studies in Hearing. Delft, The Netherlands: Delft University Press, pp. 206–209.
Viemeister NF (1970) Auditory discrimination of intensity, internal noise, and temporal processing. PhD Dissertation, Indiana University, Bloomington, IN.
Viemeister NF (1977) Temporal factors in audition: A systems analysis approach. In: Evans EF, Wilson JP (eds) Psychophysics and Physiology of Hearing. London: Academic Press, pp. 419–427.
Viemeister NF (1979) Temporal modulation transfer functions based upon modulation thresholds. J Acoust Soc Am 66:1364–1380.
Viemeister NF (1980) Adaptation of masking. In: van den Brink G, Bilson FA (eds) Psychological, Physiological and Behavioural Studies in Hearing. Delft, The Netherlands: Delft University Press, pp. 190–198.
Viemeister NF (1988) Psychophysical aspects of auditory intensity coding. In: Edelman GM, Gall WE, Cowan WM (eds) Auditory Function: Neurobiological Bases of Hearing. New York: John Wiley and Sons, pp. 213–241.
Viemeister NF, Bacon SP (1988) Intensity discrimination, increment detection, and magnitude estimation for 1-kHz tones. J Acoust Soc Am 84:172–178.
Viemeister NF, Wakefield GH (1991) Temporal integration and multiple looks. J Acoust Soc Am 90:858–865.
Wakefield GH, Viemeister NF (1984) Selective adaptation of linear frequency-modulated sweeps: Evidence for direction-specific FM channels? J Acoust Soc Am 75:1588–1592.
Wakefield GH, Viemeister NF (1985) Temporal interactions between pure tones and amplitude-modulated noise. J Acoust Soc Am 77:1535–1542.
Wakefield GH, Viemeister NF (1990) Discrimination of modulation depth of SAM noise. J Acoust Soc Am 88:1367–1373.
Wier CC, Green DM, Hafter ER, Burkhardt S (1977) Detection of a tone burst in continuous- and gated-noise maskers: Effects of signal frequency, duration, and masker level. J Acoust Soc Am 61:1298–1300.
Williams KN, Perrott DR (1972) Temporal resolution of tonal pulses. J Acoust Soc Am 51:644–647.
Yost WA, Sheft S (1989) Across critical band processing of amplitude modulated tones. J Acoust Soc Am 85:848–857.
Yost WA, Sheft S (1990) A comparison among three measures of cross-spectral processing of amplitude modulation with tonal signals. J Acoust Soc Am 87:897–900.
Yost WA, Walton J (1977) Hierarchy of masking-level differences obtained for temporal masking. J Acoust Soc Am 61:1376–1379.
Yost WA, Sheft S, Opie J (1989) Modulation interference in detection and discrimination of amplitude modulation. J Acoust Soc Am 86:2138–2147.

Zwicker E (1965a) Temporal effects in simultaneous masking by white noise bursts. J Acoust Soc Am 37:653–663.

Zwicker E (1965b) Temporal effects in simultaneous masking and loudness. J Acoust Soc Am 38:132–141.

Zwicker E (1976) Psychoacoustic equivalent of period histograms. J Acoust Soc Am 59:166–175.

Zwislocki JJ (1960) Theory of temporal auditory summation. J Acoust Soc Am 32:1046–1060.

Zwislocki JJ (1969) Temporal summation of loudness: An analysis. J Acoust Soc Am 46:431–441.

Zwislocki JJ, Hellman RP, Verrillo RT (1962) Threshold of audibility for short pulses J Acoust Soc Am 34:1648–1652.

5
Sound Localization

FREDERIC L. WIGHTMAN AND DORIS J. KISTLER

Sounds in the environment can alert, educate, entertain, or just annoy us. The information conveyed by sounds consists of time-varying streams or patterns of auditory attributes such as loudness, duration, pitch, and timbre. Extracting this information requires decoding the patterns in order to determine, for example, which attributes belong to which sound-producing objects. An important component of this source segregation process is the apparent spatial position of each object, its direction, and its distance from the listener. "Sound localization" is the term normally used to refer to the processes by which the apparent position of an object is determined.

This chapter is an admittedly biased presentation of what is known and what is not known about how sound localization works. It is not meant to be comprehensive. Rather, a small number of topics will be discussed from a somewhat different perspective than what has appeared in other recent reviews of sound localization (e.g., Middlebrooks and Green 1991). In particular, the discussion here will focus primarily on the factors governing the *apparent positions* of sounds and less on the accuracy with which listeners can indicate the actual positions of sounds or discriminate sound positions. Also, since there are so few data on the issue of the apparent distance of sound sources, there will be no discussion of this topic, even though distance is obviously an important component of apparent spatial position.

The presentation will begin with a brief history of the relevant theoretical and empirical research on sound localization. This will be followed by reviews of the basic acoustical variables and the potential acoustical and nonacoustical determinants of apparent position. Finally, after a discussion of the strengths and weaknesses of the various paradigms used to gather empirical data on sound localization, a detailed review of the empirical data will be presented.

1. Brief Historical Perspective

The lack of technology suitable for generating and controlling stimuli prevented systematic empirical study of sound localization in the 19th and early 20th centuries. Nevertheless, several theories of sound localization were proposed, attacked, and defended during this time. S.P. Thompson provided a compilation of these theories in his essay, "On the Function of the Two Ears in the Perception of Space" (Thompson 1882). As Thompson pointed out, there were basically three theoretical positions at that time. One, championed by Steinhauser and Bell, emphasized only the role of interaural level differences (ILDs), implying that interaural time differences (ITDs) were irrelevant. A.M. Mayer held a second position, arguing that both ILDs and ITDs were important. The third theory, ascribed to Mach and Lord Rayleigh, was also an ILD theory, but Mach suggested, in addition, that "... the perception of direction of sound arose from the operation of the pinnae of the ears as resonators for the highest tones to be found in the compound sounds to which the ear is usually accustomed..." (Thompson 1882, p. 410). Thompson himself favored the third position which emphasized the role played by the pinnae.

The theory formulated several decades later by Lord Rayleigh (Strutt 1907) appeared to discount the importance of Mach's pinna resonances. This later theory, which was based jointly on ITDs and ILDs, dominated the study of human sound localization for at least the next half century. Rayleigh recognized that if the acoustical wavelengths were short relative to the dimensions of the head, a listener's head would cast an acoustical "shadow" such that the level of an incoming sound wave would be greater at the ear facing the sound source than at the opposite ear, thus producing an ILD. He also noted that as a result of the different distances a sound wave would travel to the two ears there would be an ITD. In a simple experiment involving mistuned tuning forks, Rayleigh was able to demonstrate that at low frequencies, where the wavelengths were long (and the ILDs negligible), a listener's sensitivity to ITDs was greatest. Thus, he argued that localization was governed by ITDs at low frequencies and by ILDs at high frequencies. This simple theory, which came to be known as the "duplex" theory, stimulated a large number of studies in the decades that followed, most of which produced results consistent with that theory.

Much of the localization research published in the first half of this century used simple stimuli such as tones, clicks, and broadband noise and concentrated on localization in the horizontal plane. The typical results were that tones were localized less precisely than broadband sounds and that localization errors were greatest for tone frequencies around 3000 Hz (e.g., Stevens and Newman 1936). This latter finding was taken as support for the duplex theory, since 3000 Hz would be too high for effective ITD cues and too low for effective ILD cues.

The focus on simple stimuli and horizontal plane localization, and the desire for more precise stimulus control, provided the impetus for the long series of "lateralization" studies that began in the late 1940s. In lateralization experiments, headphones are used to present the stimuli so that the ITDs and ILDs can be precisely controlled. The lateralization paradigm has been used often and effectively to quantify many of the fundamental properties of binaural signal processing, such as the detectability and discriminability of ITDs and ILDs. This literature is extensive and continues to grow (see, for example, Durlach and Colburn 1978; Yost and Hafter 1987 for reviews). Unfortunately, the fact that the sound image in a lateralization experiment is consistently internalized (i.e., located inside the listener's head) suggests that the extent to which lateralization results can be generalized to issues relevant to "out-of-head" localization may be limited.

A minor revolution occurred in the field of sound localization research about 30 years ago. It appears to have been sparked by Batteau's publications in the 1960s (e.g., 1967) of his theories on the time-domain effects of the pinnae. He viewed the pinnae as highly directional reflectors that added a distinctive pattern of echoes to each incoming sound. Because of the directionality of the pinnae, the pattern of echoes would be dependent on the azimuth and elevation of the sound relative to the pinnae. Batteau's theory was that humans might decode the pattern of echoes and thus determine the direction of an incoming sound. The problem with the theory was that pinna echoes or reflections were added with very short delays, in the order of 100 to 300 μs. Thus, the temporal resolution that would be required by Batteau's decoder was nearly an order of magnitude better than current estimates of the human auditory system's temporal resolving power (Green 1971). Hebrank and Wright (1974a,b) correctly recognized that Batteau's pinna reflections would have spectral correlates and that these spectral effects might be readily decoded and used as spatial cues by human listeners. The direction-dependent filtering provided by the pinnae would thus cause changes in sound quality or timbre as a function of direction, a fact emphasized nearly 100 years earlier by Thompson (1882).

The renewed attention to pinna effects, exemplified by the work of Batteau (1967), Hebrank and Wright (1974a,b), and others, stimulated a large number of studies that have revealed the fundamental importance of the cues to sound source direction provided by the pinnae. For example, it now appears that pinna filtering is essential for localization on the vertical median plane, where interaural differences are minimal (e.g., Gardner and Gardner 1973; Hebrank and Wright 1974a,b; Butler 1975). Pinna effects also seem to be largely responsible for the externalized quality of sound images (Plenge 1974). It is also clear that pinna filtering effects are very different from one person to the next and that these individual differences can be linked to differences in localization accuracy (Belendiuk and Butler 1975; Wightman and Kistler 1989b; Wenzel, Wightman, and Kistler 1991; Wenzel et al. 1993). Pinna effects may also be responsible, in part, for our ability to discriminate between source positions in the front and rear, where interaural differences are the same.

Localization research has matured considerably in the last 10 to 15 years. First, there seems to be a recognition of the limitations of the old methods, and new methods have appeared to take their place. For example, there are fewer lateralization studies and more localization studies that probe large segments of auditory space (Oldfield and Parker 1984a,b, 1986; Wightman and Kistler 1989b, 1992; Butler, Humanksi, and Musicant 1990; Makous and Middlebrooks 1990; Kistler and Wightman 1992; Middlebrooks 1992.) New methods of stimulus generation and control (Blauert 1983; Wightman and Kistler 1989a) offer the advantages of headphone presentation without the disadvantages of internalized imaging. Second, there is greater attention paid to localization of complex, even dynamic, stimuli (Perrott and Musicant 1977; Grantham 1986) and to localization in realistic environments (Hartmann 1983; Rakerd and Hartmann 1985, 1986; Hartmann and Rakerd 1989b). Finally, while the duplex theory continues to be elaborated (e.g., Wightman and Kistler 1992), almost no study of localization is now reported without reference to the role played by the pinnae and the fundamental importance of pinna filtering cues.

2. Acoustics

The goal of most psychophysical research is an understanding of the relationships between the physical features of the stimulus and the attributes of the percept evoked by that stimulus. In many areas of hearing research, the important questions can be addressed without a complete specification of the stimulus. For example, in experiments that measure the intensity just-noticeable differences for pure tones, the acoustical stimulus is typically presented over headphones, and little attention is given to the resulting sound pressure waveform at the eardrum. The stimulus is specified in terms of the electrical waveform across the headphone terminals, since the rest of the stimulus path, consisting of headphone and ear canal, can be assumed to be linear. In localization research, however, many interesting research questions demand a more complete specification of the acoustical stimulus, often at the level of the listener's eardrums. One complication is that localization experiments are frequently conducted in a free sound field or in a treated room, so the acoustical effects of the head, torso, and pinnae must be considered. This section will present a brief review of the current state of knowledge in this area, with the emphasis on those factors that are thought to be relevant for determining apparent sound position.

It is difficult to measure the acoustical parameters of stimuli around real heads and in real ears. Moreover, given the geometrical complexity of the physical structures involved, estimating the parameters of the acoustical fields by using mathematical models is a formidable problem. For these reasons, it is convenient, as a first approximation, to model the head as a rigid sphere, with the ears represented as the endpoints of a diameter of the sphere (Woodworth 1938). This approximation produces a relatively simple mathematical model, but it ignores the obvious facts that most heads are rather elongated compared to a sphere and that the interaural axis is not a diameter of the sphere (e.g., Blauert 1983). In addition, the rigid sphere model

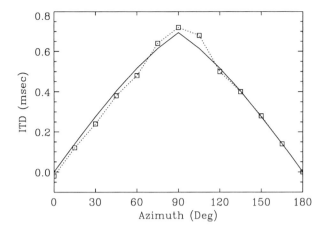

FIGURE 5.1. Interaural time difference (ITD) as a function of source azimuth for sources on the horizontal plane. The solid curve shows the prediction of the rigid sphere model, derived from plane geometry. The data points and dotted lines show the results of ear canal measurements of ITD obtained from a single subject using a wideband cross-correlation technique.

does not take into account the acoustical effects of the torso and pinnae. In spite of these weaknesses, the model generates sensible first-order approximations to the ITDs and ILDs that are thought to be important acoustical parameters for sound localization.

Initially, the ITD was estimated by noting the fact that, unless the sound source was on the median plane, the path lengths from the sound source to the two ears would be different. The path length difference, which translates into a time difference after division by the speed of sound, is readily computed using the principles of plane geometry (Woodworth 1938; Mills 1972; Green 1976; Middlebrooks and Green 1991). Figure 5.1 shows some of the ITD predictions generated by the rigid sphere model. ITD is plotted as a function of the sound source azimuth for sources on the horizontal plane. Note that the model predicts maximum ITD for a sound positioned at 90° azimuth, directly opposite one ear in the model, and a monotonic decrease of ITD with increasing departures from 90° azimuth. Also plotted in Figure 5.1 are the results of some recent measurements of ITD made by placing miniature microphones in a subject's ear canals and cross-correlating the microphone responses to wideband stimuli (for a complete description of the methods see Kistler and Wightman 1992). The close correspondence between the predictions of the rigid sphere model and actual measurements of ITD confirms the adequacy of the model for representing ITDs. Figure 5.2 shows a more complete representation of ITDs measured from several listeners. In this figure, ITD is plotted as a joint function of source azimuth and elevation with the contours of constant ITD plotted below the data.

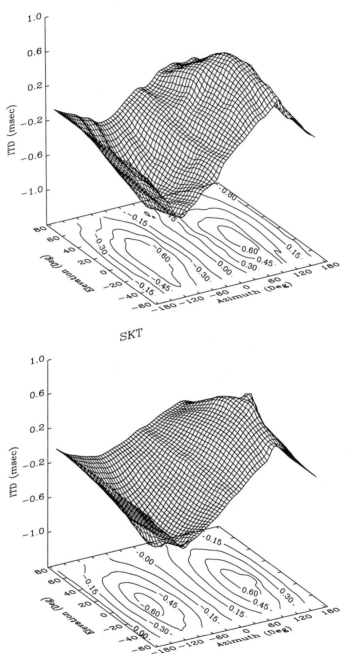

FIGURE 5.2. Measured ITDs plotted as a function of source azimuth and elevation. Measurements were taken at 15° intervals of azimuth and 12° intervals of elevation, and the data were fit with a smooth surface using a nonlinear surface fitting algorithm. The contours of constant ITD were estimated from the fitted surface. A. Subject SMC. B. Subject SKT. C. Subject SMQ. D. Subject SLV.

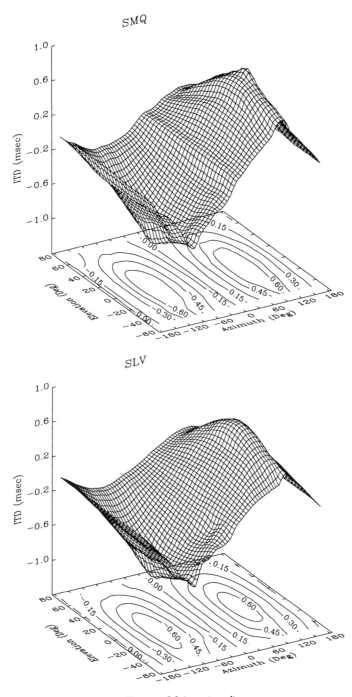

FIGURE 5.2 (*continued*)

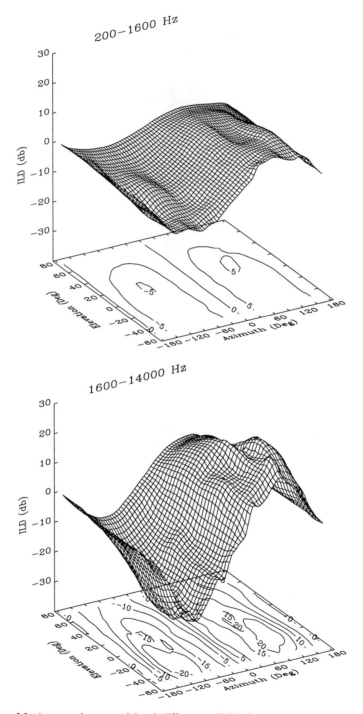

FIGURE 5.3. Average interaural level difference (ILD) from a single subject in four frequency ranges plotted as a function of source azimuth and elevation. Measurements were taken at 15° intervals of azimuth and 12° intervals of elevation, and the data were fit with a smooth surface using a nonlinear surface fitting algorithm. The

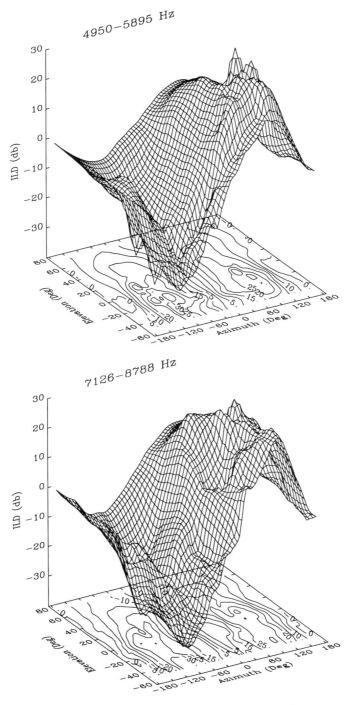

contours of constant ILD were estimated from the smooth surface. A. Frequency range is 200 to 1600 Hz. B. Frequency range is 1600 Hz to 14000 Hz. C. Frequency range is 4950 to 5895 Hz. D. Frequency range is 7126 to 8788 Hz.

The original formulations of the rigid sphere model did not acknowledge the frequency dependence of the ITD. However, Kuhn (1977) showed that a more complete form of the model predicts that ITDs are not only a function of source azimuth and elevation but also of frequency. The frequency dependence is greatest when the ITD is in the middle of its range. For sources on the horizontal plane, this would occur for a source roughly at 45 azimuth. In the extreme cases, ITDs below 600 Hz are about 1.5 times larger than ITDs above 1500 Hz.

Even an acoustical model of the head as simple as a sphere predicts that the ILD is a complicated function of both frequency and source direction (Kuhn 1977). One reason is that the magnitude of the diffracting or shadowing effect of the sphere depends on the relationship between the wavelength of the sound and the diameter of the sphere. Thus, the model predicts that ILDs will be smaller at low frequencies than at high frequencies and will vary in a complicated fashion with source direction in any narrow frequency band. At frequencies above 6 kHz, the rigid sphere model predicts ILDs as great as 15 dB (Kuhn 1987). Figure 5.3 shows measured ILDs from a single subject. In panel A, ILDs are plotted as a function of source azimuth and elevation, averaged across the frequency region 200 Hz to 1600 Hz. Panel B shows ILDs averaged across the 1600 Hz to 14000 Hz region. Contours of constant ILD are plotted below the data. Note that the low-frequency ILDs are considerably smaller than the high-frequency ILDs, as predicted by the rigid sphere model. In panels C and D are plotted ILDs in two high-frequency critical bands. Note that while the overall pattern of narrowband ILDs is similar to the pattern of wideband ILDs plotted in panels A and B, the narrowband plots reveal distinctive features that are specific to each band. The differences between wideband and narrowband ILD patterns are also reflected in the contour plots.

An essential feature of the rigid sphere model is its prediction of loci of constant ITD and ILD. The model predicts that these loci are cones that have an imaginary focus at the center of the head and an axis of symmetry coincident with the interaural axis. These are the so-called "cones of confusion" (Woodworth 1938). If only positions a constant distance from the center of the head are considered, the loci of constant interaural difference are circles, the centers of which are on the interaural axis. The ITDs and ILDs plotted in Figures 5.2 and 5.3 were measured with sources at a fixed distance from the subject. Thus, the extent to which the constant ITD and ILD contours shown in Figures 5.2 and 5.3 are roughly circular and centered approximately at 90° azimuth and 0° elevation indicates the adequacy of the rigid sphere model for representing ITDs and ILDs.

The weaknesses of the rigid sphere model have long been acknowledged. However, because of the complexity of acoustical modeling, no alternative was offered. Further understanding of the acoustics of the head and ears was delayed until actual acoustical measurements could be made of the sound field around the head and in the ear canal. The first measurements were made using dummy heads (e.g., Firestone 1930). Unfortunately, the early dummy heads did not include a complete representation of the pinnae and ear canals. More recently, a manikin has been developed (KEMAR) that is a faithful reproduction of the "average" human head, pinnae, and torso (Burkhard and Sachs 1975). This manikin is now generally accepted as a standard for acoustical measurements.

It was not until probe tubes and miniature microphones became available that it became possible to make accurate measurements in human ear canals. The classic work by Wiener and Ross (1946) and Wiener (1947) represented some of the earliest experiments in which probe tube microphones were used to measure the acoustical diffraction effects of a human head. Because these measurements were so difficult to make, the range of measurements was limited, typically including sound positions only on the horizontal plane. Ten years later, Feddersen et al. (1957) published probe tube measurements comparable to Wiener and Ross' that included, for the first time, measurements of ITD. In 1965, Shaw published the first of a long series of articles that described extensive measurements of the acoustical effects of the head, torso, and pinnae (Shaw 1965). This research effort culminated with a comprehensive summary of all the measurements then known to have been made of the pressure transformation from free field to the human eardrum (Shaw 1974) for sources primarily on the horizontal plane. A similar summary has been published by Blauert (1983). The data from all these studies clearly documented the complex acoustical effects of the head and pinnae and, in particular, how those effects depend on the frequency and direction of the sound source. Even though nearly all of the measurements were confined to positions on the horizontal plane, they revealed azimuth-dependent amplitude changes of over 20 dB (Shaw 1974) in some frequency regions.

The emergence of digital signal processing techniques and miniature microphones in the 1970s provided tools that allowed rapid measurement of both magnitude and phase of the direction-dependent transfer characteristics ofthe human external auditory system from a large number of source directions. These transfer functions have come to be known as head-related transfer functions, or HRTFs. Several researchers have published HRTF data gathered with the new techniques (e.g., Mellert, Siebrassse, and Mehrgardt 1974; Mehrgardt and Mellert 1977; Middlebrooks, Makous, and Green 1989; Wightman and Kistler 1989a; Middllebrooks and Green 1990). The most comprehensive data, obtained from 356 positions all around the subject and covering the frequency range from 3 kHz to 16 kHz, were published by Middlebrooks, Makous, and Green (1989) and Middlebrooks and Green (1990).

The recent data reveal that the acoustical transformation from free field to the eardrum is more complicated than that predicted by the rigid sphere model. This is primarily a result of the highly directional filtering action of

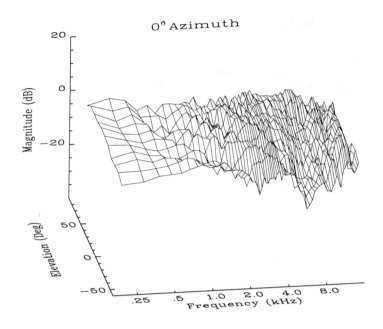

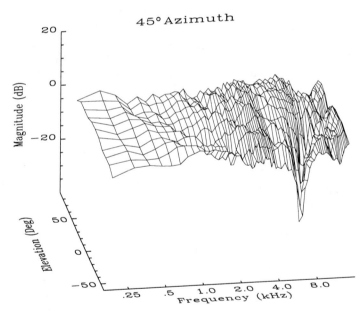

FIGURE 5.4. Plot of the directional transfer functions (DTFs) measured from a single subject for sources at three azimuths and elevations from −48° to 54° at 12° intervals. A. 0° azimuth. B. 45° azimuth. C. 90° azimuth.

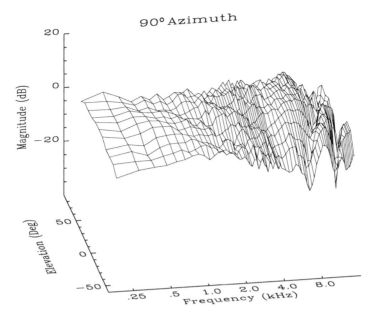

FIGURE 5.4 (*continued*)

the pinnae. Figure 5.4 illustrates this directional filtering by plotting the "directional transfer functions" measured from one ear of one subject for sources at several different azimuths and elevations (see Kistler and Wightman 1992 for a description of the methods). Directional transfer functions (DTFs) are functions that represent departures from the diffuse field response of the ear. The diffuse field response is estimated by computing the root-mean-square of the HRTF magnitudes from all directions measured. The DTFs represent magnitude differences (in dB) between HRTFs and the diffuse field response. Thus, the DTFs show only the directional effects. Note that, especially at high frequencies, there are sharp peaks and valleys in the DTFs that change character dramatically with changes in source direction. Several investigators have singled out for special consideration the deep spectral notch in the HRTF between 5 kHz and 10 kHz (see Fig. 5.4C) that appears to move systematically in frequency as the elevation of the source is changed (e.g., Butler and Belendiuk 1977; Neti, Young, and Schneider 1992). However, this notch is only one of many highly directional spectral features that characterize measured HRTFs.

The recent measurements also reveal large individual differences in HRTFs. In their study of 10 listeners, Wightman and Kistler (1989a) reported that in the frequency range from 5 kHz to 10 kHz, where the pinna contributions are especially prominent, the HRTFs from individual listeners can be expected to differ by as much as 28 dB. Until the significance of these individual differences is understood, interpretations of HRTF measurements made with "average" dummy heads must be made with some caution.

Kistler and Wightman (1992) have recently proposed a statistical model of HRTFs that accounts for more than 90% of the variance across both source positions and listeners in HRTF measurements. The model is based on a principal components analysis (PCA) of the DTFs from a group of 10 listeners and 265 source positions. The PCA revealed that HRTF magnitude functions can be represented by a weighted sum of five basic spectral shapes. The weights are specific to the subject and source position. Inspection of the basic spectral shapes from which all the HRTFs can be derived shows that most of the variance in HRTF shape is at frequencies above 5 kHz and is thus most likely a result of pinna diffraction effects. Using different listeners and different techniques to measure HRTFs, Middlebrooks and Green (1992) obtained nearly identical basic spectral shapes from a PCA of their HRTF magnitude data.

The complexities apparent in measured HRTFs highlight the limitations of the rigid sphere model. While that model provides a reasonable first approximation to measured ITDs, it is clearly inadequate to explain the amplitude directionality of HRTFs, much of which is provided by the pinnae. The relative importance of the ITDs and the amplitude directionality as determinants of apparent sound position will be discussed in Section 3.

3. Potential Determinants of Apparent Sound Position

While nearly all systematic research on sound localization has emphasized the importance of acoustical cues for determining apparent sound position, it now seems clear that sound localization involves the integration of information from a wider range of sources. Thus, in addition to acoustics, other factors such as visual cues, memory, and a listener's expectations are now thought to have significant influence on the apparent position of an auditory object. In some situations, such as those involving simultaneous auditory and visual stimulation, the nonacoustical cues can even override the acoustical cues. Unfortunately, there are relatively few data on these nonacoustical factors, so the emphasis in this section will be on the acoustical cues used by humans to localize auditory objects.

3.1 Interaural Differences

The fact that human beings have two ears leads naturally to the hypothesis that localization of sound sources involves some kind of interaural comparison of encoded acoustical information. Indeed, computation of either interaural differences or cross-correlation is the primary basis of nearly all major theories of localization, including not only the classical duplex theory but also the more modern theories (Colburn 1973; Searle et al. 1976; Stern and

Colburn 1978; Lindemann 1986). As Section 2 showed, HRTF measurements reveal substantial ITDs and ILDs that are highly dependent on the direction of a sound source. As evident in Figure 5.1, the ITD grows from near zero for sources on the vertical median plane to almost 700 μs for a source on the horizontal plane at 90° azimuth. Given that ITDs of less than 10 μs can be detected (Klumpp and Eady 1956), there is little doubt that ITD could be an effective cue to source position. Interaural level differences (ILDs) as large as 30 dB have been measured at high frequencies for sources well off the vertical median plane (Middlebrooks, Makous, and Green 1989). The detection threshold for ILD is less than 1 dB (Mills 1960), so this cue as well can be considered potentially effective for signaling apparent sound source position.

While both ITDs and ILDs are plausible determinants of apparent position, there are factors that limit the potential effectiveness of each. For example, ITDs are likely to be most effective if the sound contains low frequencies or has a low-frequency envelope. There are at least two reasons for this. First, coding of temporal information by the auditory system is most precise at low frequencies (e.g., Yin and Chan 1988; Joris and Yin 1992). Second, at low frequencies the sound levels at the two ears will be roughly equal, since ILDs are smallest in this frequency region. A large level imbalance between the ears could lead to errors or bias in coding the ITD.

To the extent that the ITD alone signals apparent sound image position, large discrepancies can be expected between apparent position and actual position because the ITD is an ambiguous cue for position. Recall that the ITD is constant for all sources that lie on a cone of confusion. If distance is fixed, the ITD is constant for sources on a circle centered approximately on the interaural axis. Figure 5.2 illustrates the approximately circular loci of a constant ITD measured from a typical subject. It must be expected that, if listeners rely only on the ITD, source positions on these constant ITD contours would not be discriminated and that there would be frequent localization errors. For example, for sources on the horizontal plane, a position in the front, at 30° azimuth, would be confused with the complementary position in the back, at 150° azimuth. The fact that such front-back confusions occur infrequently in realistic listening conditions suggests that apparent position is determined by more than ITDs alone.

Interaural level difference can be expected to be most salient as a position cue at high frequencies. In the low frequency region, below about 1600 Hz, the wavelengths of sounds are long compared to the dimensions of the head, and the ILDs are relatively small. Figure 5.3 illustrates this point. Note that while ILDs are generally smaller in the low frequencies (Fig. 5.3A) than in the high frequencies (Fig. 5.3B), the low-frequency ILDs at some positions are sufficiently large (5 dB) to be considered potential position cues. Note also that when ILDs are averaged across a broad frequency region, as in Figure 5.3, the contours of constant ILD are reminiscent of the constant ITD contours (Fig. 5.2). Thus, the broadband ILD would be just as ambiguous a position cue as the ITD. The broadband ILD can indicate only the extent to which a source lies to the right of the median plane (positive azimuths in Fig. 5.3) or to the left (negative azimuths in Fig. 5.3). The elevation of the source and whether it is in front or in the rear cannot be uniquely coded by the broadband ILD. In any narrow frequency band, a given ILD is similarly ambiguous, in that it indicates a locus of potential positions (Middlebrooks, Makous, and Green 1989; Fig. 5.3C,D). The ILD can be a reliable indicator of source position only if the listener analyzes the pattern of ILDs in several frequency regions. The pattern of ILDs across the high-frequency regions changes in complicated ways as the position of a sound changes, as exemplified by the differences between Figures 5.3C and 3D. While some researchers (Butler and Humanski, 1992, Searle et al. 1976) have argued for the importance of the systematic change in one feature of this pattern, such systematic changes typically occur only within a limited range of source azimuth and elevation. Thus, it appears unlikely that a single feature of the ILD patterns can serve as a reliable spatial cue. It is more plausible that some subset of the features in the entire pattern is extracted and compared to remembered templates in order to determine source position. Some evidence that this may occur comes from studies in which one listener is presented the pattern of ILDs from another listener (Butler and Belendiuk 1977; Wenzel, Wightman, and Kistler 1991). Since the patterns of ILDs vary considerably from one listener to another (Wightman and Kistler 1989a), the fact that listeners in these experiments reported slightly different apparent positions when listening with different ILD patterns can be taken as preliminary evidence supporting the notion that patterns of ILDs are processed in an idiosyncratic manner.

3.2 Monaural Spectral Cues

The head and pinnae act as filters to modify the spectrum of an incoming sound. This filtering not only creates the direction-dependent interaural differences discussed in Section 3.1 but also changes the spectrum of the sound at each ear individually. The directional effects superimposed on the spectrum of a sound at each ear by the outer ear filtering are called "monaural spectral cues." It has often been suggested that these monaural spectral cues might serve as a basis for apparent sound position. That point of view is supported by a substantial body of data which suggests that monaural listeners, who are presumed not to have access to normal interaural difference cues, can localize sounds with an accuracy well above chance (e.g., Angell and Fite 1901; Oldfield and Parker 1986; Butler, Humanski, and Musicant 1990).

A logical problem with the notion that monaural spectral cues might signal sound position is that the spectrum received at the ear is the product of the monaural filtering caused by the head and pinna and the spectrum of the sound source itself. Thus, the only way the two components of the product could be deconvolved, in order to extract the monaural position cue, is through knowledge of the spectrum of the source. This fact was clearly acknowledged by the first proponents of theories of localization based on monaural spectral cues. For example, Thompson (1882) writes: "The ear has been trained from childhood to associate certain differences in the quality of sounds (arising from differences in the relative intensities of some of the partial tones that may be present) with definite directions; and, relying on these associated experiences, judgments are drawn concerning sensations of sound whose direction is otherwise unknown. For sounds that are familiar a difference of quality as heard *in one ear* will at once suggest a direction. It is completely open to doubt whether a pure simple tone heard in one ear could suggest any direction at all." (1882, p. 415) Thus, the only conditions under which the monaural spectral cues could serve as a reliable indicator of source position are those in which the source spectrum is known in advance. Nearly all of the studies of localization by monaural listeners cited above used either deterministic or stationary stimuli, the spectra of which were both relatively smooth over a broad frequency range and unchanging from presentation to presentation. It is possible that the results of these experiments would indicate less reliance on monaural spectral cues if stimuli with more uncertain or irregular spectra were used. It also seems likely that, in daily life, listeners come to learn the basic spectral characteristics of some familiar sounds, so that, in spite of the logical problems, monaural spectral cues may play an impoltant role in everyday sound localization.

3.3 Dynamic Cues

An essential feature of all the potential acoustical determinants of apparent sound position is that, in everyday life, they are constantly changing. The acoustical cues vary both with changes in the position of the sound source and with changes in the position of the listener's head. Given a fixed source position, movement of a listener's head changes the angular relation between the head and the source, thus changing both the interaural differences and the monaural spectral cues. There has been considerable speculation and some research on the salience of these dynamic cues for determining apparent sound position. Some have argued, for example, that front-back confusions could be resolved by noting the direction of change in the ITD following a rotational movement of the head around the vertical axis (Wallach 1940). Given that the left-right position of the source and the direction of the head movement are known, an increase or decrease in the ITD would unambiguously indicate whether the source was in the front or the rear. A simple extension of this argument has shown that head movements can resolve position ambiguities anywhere on the "cone of confusion" (Wallach 1939). Head movements might also account for the success of monaural localizers who are presumed to rely on monaural spectral cues to sound source position. Recall that the effectiveness of a monaural spectral cue depends logically on the extent to which the listener knows the spectrum of the source in advance. This dependence could theoretically be avoided if the listener could sample the sound source twice, at two different head positions. The change in the spectral content of the samples would not be dependent on the spectrum of the source.

The results of several experiments suggest that head movements play a rather minor role in determining apparent sound position. Thurlow and Runge (1967) showed that induced head movements caused a small drop in localization error, but only in some of the conditions they studied. Pollack and Rose (1967) also reported a very small effect of head movements in their study of localization on the horizontal plane. Fisher and Freedman (1968) failed to show any effect of head movements. Thus, in spite of the compelling logic of Wallach's initial statements, the available empirical evidence does not support the view that head movements play a significant role in determining apparent sound position.

3.4 Nonacoustical Factors

It is clear that, under certain conditions, the apparent position of a sound source can be affected significantly by nonacoustical factors. Vision exerts the most widely recognized nonacoustical influence on apparent sound position. The "ventriloquism effect" is an example of how the presence of a corresponding visual object can bias judgments of the apparent position of an auditory object (Jackson 1953; Pick, Warren, and Hay 1969; Warren 1970). There is also some evidence that the mere presence of a visual environment can influence apparent position, since localization accuracy is slightly greater in the light than in the dark (Shelton and Searle 1980). It is probably because of the powerful influence of visual cues that experiments on apparent sound position are most often conducted under conditions in which the listeners cannot see the actual sound sources.

A listener's familiarity with the characteristics of a sound source is a second nonacoustical factor that probably plays a role in determining apparent spatial position. There has been some speculation, for example, that this factor is involved in signaling the apparent distance of a source (Coleman 1962; Gardner 1969). The idea is that, as distance increases, there is a corresponding increase in the high-frequency attenuation caused by atmospheric scattering. This effect could serve as a cue to distance if the listener knew the spectral characteristics of the source itself. The argument here is similar to that made above in connection with monaural spectral cues. Familiarity, resulting from long-term exposure to specific sources (e.g, the human voice), might provide the necessary a priori knowledge of the stimulus spectrum.

The stimulus and response context in which a sound is presented contributes significantly to a listener's judgment of apparent position, if not to apparent position itself. There are several examples of this effect in the literature from experiments in which the number of sources and potential responses made available to the listener were small. In the experiment reported by Butler and Planert (1976), for example, five sources were arranged in the vertical median plane, 15° apart in one condition and 7.5° apart in another condition. The listener's task was to indicate which of the sources actually produced the stimulus. The number of sources, the number of stimuli presented from each source, and the number of response alternatives were the same in the two conditions. The distributions of responses from the seven listeners were also nearly identical in the two conditions, in spite of the fact that the sources at the extremes were in positions 16° apart in the two conditions. Given the fact that the stimuli were bursts of bandpassed noise, 2 kHz wide and centered at 8 kHz, it can probably be assumed that the apparent positions of the stimuli in the two conditions were actually different, even though the distributions of responses were the same. Another example of the strong influence of context can be found in the experiment reported by Butler and Humanski (1992). In separate conditions of this experiment, noise stimuli were presented from seven sources arranged either in the vertical median plane or in the lateral vertical plane (the vertical plane containing the interaural axis). One of the source positions, namely that at 90° elevation or directly overhead, was the same in the two conditions. As in the Butler and Planert (1976) study, the task was to identify the loudspeaker that actually produced the sound. For the purpose of this discussion, the important result of this experiment was the fact that the listeners' judgments of the elevation of the overhead source were very different in the two conditions, in spite of the fact that the source position was the same. The mean data (Butler and Humanski 1992, Fig. 2) indicated a reported elevation of about 45° in the median plane condition and near 90° in the lateral plane condition. Whether such effects reflect response biases or perceptual effects is, of course, unknown.

4. Research Questions and Experimental Paradigms

Human sound localization can be studied from both an objective and a subjective perspective. Objective localization refers to the accuracy with which listeners can identify the spatial position of a real sound source or discriminate among sounds on the basis of position. Subjective localization, in contrast, refers only to the apparent position of an auditory object and not to the correspondence between apparent position and actual position. The distinction between objective and subjective localization is useful since different experimental questions are posed, different paradigms are used, and different interpretations are made of experimental data, depending on the perspective. Generalizing research results from one domain to the other is not always straightforward. For example, experiments that examine the factors governing the just detectable change in the angular position of a sound are objective localization experiments. These experiments, called "minimum audible angle" or MAA experiments, contribute valuable knowledge about binaural processing, but they provide little or no information about the factors responsible for the apparent positions of the sounds. A discrimination paradigm is typically used to measure the MAA, and listeners might discriminate among the sounds using cues such as subtle spectral differences that may or may not be relevant to apparent position. The distinction is the same as that between frequency discrimination and pitch. Frequency discrimination is objective, while pitch perception is subjective. Results from frequency discrimination experiments cannot always be generalized to the pitch domain, since frequency discrimination and pitch perception are mediated by auditory processes that are probably only remotely related.

Generalizing results from subjective localization experiments to issues of objective localization can also be misleading. As an example, consider an experiment in which one ear of each listener is blocked, rendering the listeners "monaural" (e.g., Hebrank and Wright 1974a). Loudspeakers are arranged on the horizontal plane on the unoccluded side, and on each trial a sound is presented from one of the loudspeakers. Listeners are asked to indicate the apparent position of the sounds. As frequently occurs in experiments like this (e.g., Hebrank and Wright 1974a; Butler 1975) the listeners' responses are strongly biased toward the 90° azimuth position, directly opposite the unoccluded ear. In other words, regardless of the actual position of the sound source, the apparent position was nearly the same, at 90° azimuth. Expressed in objective terms, these results would show that error was high for sources near the median plane (0° or 180° azimuth) and low for sources near 90° azimuth. The error would progressively decrease as the source azimuth approached 90°, so it would be tempting to conclude that, with one ear blocked, horizontal localization is near normal for sources at 90° azimuth and substantially degraded only near the median plane. This would be a reasonable conclusion if only the accuracy of localization were considered. However, from the point of view of the apparent position of the sounds and the factors governing apparent position, the conclusion would be that localization is poor in the monaural condition since large changes in actual source position cause only small changes in apparent position.

Research on human sound localization is technically demanding. One problem is that the stimulus is difficult to measure and to control. In the acoustical settings that seem most relevant to human localization (e.g., noisy rooms), the pressure waveforms that reach a listener's ears from even the simplest sound sources are "contaminated" by reflection and diffraction effects. Even minute changes in the direction of the source or the point of measurement of the waveform have major consequences. Systematic manipulation and control of the stimulus under these conditions is problematic at best. An additional problem is that indirect techniques must be used to quantify the subjective aspects of localization such as the apparent position of an auditory image. The characteristics of these measurement tools are not well documented or understood. In response to these technical challenges, researchers have used several different paradigms to study sound localization, each with a unique set of strengths and weaknesses. No single technique has emerged as the most appropriate for addressing all of the research questions.

The paradigms most often used in sound localization research can be classified according to both stimulus presentation technique and psychophysical procedure. There are two stimulus delivery techniques: headphones and loudspeakers, and three psychophysical procedures: discrimination, categorization, and absolute judgment, producing six potential paradigms. Only five of the six have actually been used.

Consider first the paradigms which involve presentation of stimuli with headphones. Headphone stimulus presentation has the advantage of allowing more precise manipulation and control of ITDs and ILDs than is feasible with loudspeaker presentation. The headphone experiments are described as involving "lateralization," as opposed to "localization," because the resulting sound image is reported to be inside the head, usually somewhere on a line between the ears. Lateralization paradigms have been used to address both objective and subjective issues related to sound localization. For example, the detectability and discriminability of ITDs and ILDs have been measured by using discrimination procedures in a lateralization paradigm. Discrimination procedures, as the name implies, require a listener to make a choice between sounds. Since the dependent measure in a discrimination experiment is some measure of accuracy, like percent correct, this experiment quantifies an objective aspect of sound localization.

Lateralization experiments have also been used to study subjective issues, such as the effects of manipulations of interaural differences on apparent image position. For example, several studies have used an absolute judgment procedure to show that an interaural time difference of 300 μs or an interaural level difference of about 5 dB is sufficient to move a lateralized image about halfway from the center to the side of the head. In an absolute judgment procedure, listeners estimate the apparent position of the sound directly. In lateralization experiments, this is typically done by having listeners indicate the extent to which the sound image is "lateralized." A response of 100%, for example, might mean that the image is as far to one side as it can go. The only problem with the use of lateralization experiments to study the factors that govern apparent sound image position is that the apparent position of a lateralized image is not easily related to the apparent azimuth, elevation, and distance of an externalized image. Thus, the generalization of lateralization results to localization conditions is not always straightforward.

A wide variety of questions about both objective and subjective aspects of sound localization has been addressed in experiments which involve the presentation of stimuli over loudspeakers. The MAA experiment is an example of one that measures an objective aspect of localization, the "just-noticeable" shift in the position of a sound source (Mills 1958). Note that the MAA experiment assesses only the discriminability of sounds presented from two positions and does not quantify the apparent positions of those sounds, the subjective attributes. The categorization and absolute judgment procedures, in contrast, produce data directly relevant to the apparent positions of sound produced by loudspeakers.

The term "categorization" is used to refer to procedures in which listeners are given a small number of response alternatives with which they register their judgment of apparent sound image position. For example, in many categorization experiments, the listener is given a small set of response alternatives (usually 5 to 9) on the horizontal plane or on the vertical median plane (e.g., Butler 1969; Blauert 1971; Hebrank and Wright 1974a,b; Belendiuk and Butler 1975; Butler and Planert 1976; Butler and Flannery 1980; Hartmann 1983; Musicant and Butler 1985). Then, after each stimulus presentation, the listener indicates which available position most closely matches the apparent position of the sound image. One serious problem with the categorization paradigm is that if the apparent position of a stimulus is beyond the range of the response alternatives, the listener is forced to make a judgment that does not actually represent apparent position. This problem is particularly evident in experiments on monaural localization of sounds presented in the vertical median plane (e.g., Hebrank and Wright 1974a). In these experiments, the sound sources and response alternatives are arranged on the vertical median plane. However, because one ear is blocked, the apparent positions of the sounds are off the median plane, toward the unoccluded ear. Because listeners are forced to use only the available response alternatives, the data nominally indicate apparent positions on the median plane, while the actual apparent positions of the sounds were not on the median plane.

Other problems with the categorization paradigm result from context effects (Parducci 1963). For example, if the number of response alternatives is small (5 to 9) or if the differences among the stimuli are subtle, listeners may distribute their responses among the alternatives in ways that depend more on the distribution of stimuli than on the distribution of apparent positions. In other words, different sets of stimuli that actually have different apparent positions may produce the same distribution of responses in categorization paradigms (e.g., Butler and Planert 1976). Moreover, identical stimuli may be categorized very differently if the context of stimuli is changed (Butler and Humanski 1992). The fact that most localization studies that used the categorization paradigm have used seven or fewer response alternatives evenly spaced on either the horizontal or vertical median plane must be taken into consideration when evaluating the results of these studies.

For measuring the apparent position of a sound presented by a loudspeaker, the absolute judgment procedures offer some significant advantages. Absolute judgment requires a listener to estimate the apparent position of an auditory image directly, with a more or less continuous scale and with the response relatively unconstrained. In typical absolute judgment experiments, sounds are presented from a large number of positions all around the listener. Several variations of the absolute judgment procedure have appeared in the past ten years. In one, a stimulus is presented to blindfolded listeners who then point to the apparent position with a hand-held device similar to a pistol (Oldfield and Parker 1984a,b, 1986). The orientation of this "gun" is then recorded photographically. Apparent distance can not be measured with this technique. In a similar procedure, a listener points his/her nose toward the auditory image, and the orientation of the head is registered with a magnetic tracking device (Makous and Middlebrooks 1990; Middlebrooks 1992). A third absolute judgment procedure is that used by Wightman and his associates (Wightman and Kistler 1989b, 1992; Kistler and Wightman 1992), in which listeners call out numerical estimates of apparent azimuth, elevation, and distance. All three procedures produce comparable data when listeners judge the apparent positions of wideband sounds presented in a free field. The major difference among them is in the variance of the responses. The "nose-pointing" technique seems to produce data with the least variance, and the verbal technique produces the most variable data. The main problem with some absolute judgment procedures is that it is difficult to quantify how much of the variance in the responses is "perceptual" (i.e., a result of image diffuseness or position uncertainty) and how much can be attributed to components of the response process, such as familiarity with the coordinate system or motor ability. Nevertheless, the absolute judgment paradigms seem uniquely capable of providing data that are easily interpreted as measures of such subjective attributes as apparent sound image position.

5. Experimental Data on Apparent Sound Position

This section summarizes the empirical results that reveal the relative salience of the potential determinants of apparent sound position that were described in Section 3. Since several comprehensive reviews of this literature have been published during the last decade (Blauert 1983; Yost and Gourevitch 1987; Middlebrooks and Green 1991), the discussion here will concentrate on new developments or perspectives and on results not covered in those previous reviews.

The major acoustical determinants of apparent sound position, interaural differences and monaural spectral cues, might be described as "wideband" cues, since changes in sound position cause differential changes in the values of these parameters across the frequency spectrum. Interaural time difference, for example, can be as much as 50% greater at low frequencies than at high frequencies (Kuhn 1977). Interaural level difference is also frequency dependent but in a more complicated fashion. In addition to the ILD created by a head shadow, which is much greater at high frequencies than at low frequencies (see Fig. 3), there is the spectral patterning created by pinna filtering. The result is a pattern of ILDs across frequency that is both complex (numerous peaks and dips) and rapidly changing with source direction. Monaural spectral cues, also a product of pinna filtering, are similarly wideband and complex.

While the major cues are wideband, much of the research on apparent sound position has used relatively narrowband stimuli such as tones and filtered noises. There are several good reasons for this. For one, the acoustics of narrowband stimuli are generally easier to measure and control. For another, the duplex theory makes different predictions about how low- and high-frequency sounds are localized. However, when considering the results of experiments on narrowband sounds, it is important to recognize that the available cues to sound position are quite limited with these stimuli. Thus, it is possible that the apparent positions of such sounds are determined quite differently than when the entire constellation of cues is available.

Similar reservations can be raised about the results of experiments in which cues are distorted or are presented in unnatural combinations. One example is the sizeable group of experiments on "monaural" localization, in which normal hearing listeners are rendered monaural by plugging one ear. Users of this paradigm usually describe it as eliminating interaural difference cues. However, since the binaural neural pathways are intact and functioning even with one ear covered, it may be more appropriate to view the "monaural" paradigm as involving distorted, or at least unnatural, interaural difference cues. The cues are unnatural because, for example, the ear plug creates large ILDs even at low frequencies. Another example of a paradigm involving unnatural or distorted cues is the lateralization paradigm, in which stimuli are presented over headphones. Headphone-presented stimuli bypass the pinnae and thus do not contain normal monaural spectral cues. Both the "monaural" and lateralization paradigms have produced valuable information about binaural processing, information that is clearly relevant to our understanding of the determinants of apparent sound position. However, as was mentioned in Section 4, the extent to which results from those paradigms can be readily generalized to more complex, natural listening situations is an open question.

5.1 Narrowband Sounds

Sounds with an effective bandwidth of less than a critical band contain impoverished directional cues. The ITD and ILD cues are available only in one frequency band, thus effectively eliminating cross-frequency pattern analysis. Monaural spectral cues are also virtually eliminated. Narrowband ITDs and ILDs do not unambiguously cue a single sound source position because of the cones of confusion on which ITDs and ILDs are nearly constant (see Figs. 2 and 3). Thus, it is not surprising that narrowband sounds are often difficult to localize.

Pure tones are especially hard to localize, since interaural phase leads and lags are indistinguishable with periodic stimuli. In other words, given two sinusoids A and B with period T seconds and with a t second delay between them, there is no way to know whether A leads B by t seconds or B leads A by $T - t$ seconds. At low frequencies, there is little confusion because the period of the stimulus is long compared to the maximum expected ITD. However, as the frequency increases, a phase lead, which might normally indicate an ITD produced by a source to the right side, is more easily confused with a phase lag, which would signal a source on the left. Consider a 1000-Hz tone presented from a source directly opposite the right ear of a listener. This situation would produce an ITD of approximately 700 μs with the right ear signal leading. Since the period of the 1000-Hz tone is 1000 μs, the 700-μs right-leading ITD cannot be distinguished from a 300-μs left-leading ITD, which would normally cue a source approximately 50° to the left of the midline. The problem is even more serious at a higher frequency where the period of the sinusoid is shorter.

The classic studies reported by Stevens and Newman (1936) and Sandel et al. (1955) document several of the problems with localization of narrowband stimuli. In both of these experiments, listeners used an absolute judgment procedure to report the apparent positions of tonal stimuli presented from loudspeakers on the horizontal plane. The results are characterized by large localization errors, as great as 20° for tone frequencies between 1000 Hz and 3000 Hz. In addition, Stevens and Newman reported a high frequency of front-back confusions. These errors are expected given the impoverished and ambiguous directional cues available in narrowband stimuli.

The pattern of errors in pure-tone localization studies contrasts sharply with measurements of the precision of sine wave localization obtained from the MAA paradigm (Mills 1958; Hartmann and Rakerd 1989a). The MAA is between 1° and 3° on the horizontal plane, depending on frequency, and increases rapidly as the source azimuth approaches 90°. The average pure-tone localization error is nearly an order of magnitude greater and shows little dependence on azimuth. However, the MAA paradigm requires listeners only to *discriminate* between two sounds and not necessarily to judge their apparent positions. Since the MAA paradigm involves detection or discrimination, it can only provide information regarding the sensitivity of the binaural system to subtle differences between sounds. The information it contributes about the determinants of apparent source position is, therefore, limited. Moreover, since the MAA and apparent position paradigms are very different, there is no guarantee that listeners are attending to the same stimulus attributes in each paradigm. Thus, a lack of correspondence between MAA and apparent position data is not surprising.

Narrowband sounds presented on the vertical median plane offer virtually no directional cues. Interaural differences are minimal, and, since the sounds are narrowband, monaural spectral cues are virtually nonexistent. When listeners are asked to indicate the apparent positions of such sounds, the responses indicate little or no correspondence between actual position and apparent position. Rather, it appears that the frequency of the sound determines its apparent position (Roffler and Butler 1968; Blauert 1969). The early work on this problem indicated that higher frequency tones are judged to have higher positions (Pratt 1930; Trimble 1934; Roffler and Butler 1968). Blauert conducted extensive followup studies in the 1960s. The results, summarized in his book Blauert (1983, pp. 107–108), indicated that, with narrowband stimuli, the relationship between stimulus frequency content and apparent elevation is more complicated and probably linked to pinna filtering effects. For example, Blauert (1969) reported that a narrowband noise centered at 8 kHz had an apparent position overhead and that the 8-kHz region was amplified by the pinnae for sources overhead. Blauert's "directional bands," Butler's "covert peaks" (Butler, Humanski, and Musicant 1990), and various other theories that emphasized the importance of spectral peaks and notches (e.g., Hebrank and Wright 1974b; Bloom 1977; Butler and Belendiuk 1977; Watkins 1978) are all manifestations of the same basic hypothesis: that the apparent elevations of narrowband sounds are governed by pinna filtering effects.

The most comprehensive study to date on the localization of narrowband sounds was recently published by Middlebrooks (1992). The paradigm used by Middlebrooks differed in a number of ways from those used by his predecessors. The most notable difference was that neither the sound sources nor the listener's judgments were restricted to the horizontal or vertical planes. High frequency (6 kHz and above) narrowband noises were presented from a large number of positions all around the listener, and listeners indicated apparent position with an unconstrained "nose-pointing" procedure. In addition, Middlebrooks measured, in considerable detail, the acoustical properties of each listener's outer ears, so that precise estimates could be made of the stimulus at each eardrum. The results indicated that the apparent source azimuth was generally quite close to the actual source azimuth and probably determined by ILDs. Apparent source elevation, however, was governed entirely by frequency content, not only for sources on the vertical median plane, but also for sources in other regions. An analysis of the data suggested that the listeners perceived each sound as originating from an azimuth given by the ILD cue and from an elevation at which the pinna filter functions associated with that position most closely matched the eardrum stimulus spectra. The pinna filter functions were assumed to be recorded in the listener's memory. Middlebrooks embodied the basic elements of this spectral pattern matching scheme in a quantitative model in which the contributions of ILDs and spectral cues to the listener's position judgments were estimated separately and combined. The model was tested using techniques borrowed from signal detection theory. On the basis of these tests, Middlebrooks concluded that the ILD cues and spectral shape cues make independent contributions to the apparent positions of the narrowband sounds used in his experiments. While the extent to which these results might generalize to the apparent positions of broadband or low-frequency sounds is not clear, Middlebrooks' elegant experiment leaves little doubt that, under certain conditions, the spectral cues provided by the pinnae can be very potent.

Lateralization experiments have provided important information about the determinants of the apparent position of narrowband sounds. Since the ITD and ILD can be independently manipulated in a lateralization experiment, the paradigm can be used to estimate the relative salience of the two cues. Blauert (1982), for example, measured the intracranial position of narrowband noise stimuli with a frequency-independent ITD and zero ILD. Note that, with an ILD of zero, which would signal a source on the median plane, the stimuli contained conflicting interaural cues. The data suggested that for each ITD the internalized image was shifted to the side by a constant amount for band center frequencies below about 1.5 kHz and by a much smaller amount for band center frequencies above that. At high frequencies, the internalized image was centered, the position indicated by the zero ILD cue. Trahiotis and Bernstein (1986) reported data from a comparable experiment that are at least qualitatively similar to those reported by Blauert (1982). The results of both studies suggest that ITD is more salient than ILD as a localization or lateralization cue at low frequencies and ILD is more salient at high frequencies.

5.2 Broadband Sounds

As Section 5.1 illustrates, the apparent positions of narrowband sounds often do not even come close to the actual position of the sound source. The reason for this is that no single frequency band can convey an unambiguous cue to source position. A single-band ITD and ILD combination always signals a *locus* of potential source positions, the "cone of confusion." If a sound has a bandwidth larger than a single critical band, the directional cues provided by the individual bands can be compared and combined. In other words, a wideband sound offers the possibility of cross-spectral comparisons of directional cues, which might resolve the directional ambiguities associated with narrowband sounds. For example, for sources restricted to the horizontal plane, locations in the front might be distinguished from locations in the rear by virtue of the fact that frontal locations have a slight high-frequency emphasis compared to rear locations, as a result of pinna filtering. Thus, so long as the sound contains sufficient high-frequency energy, this monaural spectral cue might be used to resolve the "cone of confusion" ambiguity. Of course, with a wideband sound, the ILD cue is likely *never* to be ambiguous, since the ILD is represented by a pattern of ILDs across frequency, rather than a single value. This ILD pattern could also aid the resolution of ambiguities such as front-back confusions.

The correspondence between actual and apparent sound source position increases as the bandwidth of the stimulus is increased, implying that listeners do integrate cues across frequency bands, as suggested above. For example, Butler (1986) reported that with bandpass noise stimuli (8-kHz center frequency) on the horizontal plane there was a considerable increase in localization accuracy as the stimulus bandwidth was increased from 2 kHz to 8 kHz, with the most notable effect being a reduction in front-back confusions. For sources on the vertical median plane, the bandwidth effect appears to be even greater, as might be expected since ITD and ILD cues are minimal on the median plane. Butler and Planert (1976) reported nearly a ten-fold reduction in error as the bandwidth of the bandpass noise stimulus (centered at 8 kHz) is increased from 1 kHz to over 15 kHz. The broader bandwidth stimuli on the vertical median plane clearly provide the monaural spectral cues which are absent in narrowband stimuli and which are essential cues in the absence of ITDs and ILDs. The strong frequency-dependent biases evident in the apparent position data reported by Middlebrooks (1992) disappear when the bandwidth of the sounds is increased. Recall that with narrowband sounds Middlebrooks' listeners reported apparent azimuths roughly coincident with actual azimuth, but apparent elevations that were determined by band frequency. With wideband sounds, both the apparent azimuth and apparent elevation coincided with the actual azimuth and elevation, implying that cues were being integrated across frequency bands in the wideband condition.

As a consequence of cue integration across frequency, localization of wideband sounds is, in general, much more precise than localization of narrowband sounds and involves few confusions and biases. Support for this conclusion comes from four studies of the apparent positions of wideband sounds that have been published in the last few years (Oldfield and Parker 1984a; Wightman and Kistler 1989b; Butler, Humanski, and Musicant 1990; Makous and Middlebrooks 1990). The data from both the Makous and Middlebrooks study and the Oldfield and Parker study suggest that error is less than 5° for sources in front at ear level, slightly larger for sources at the side and above and below the horizontal plane, and almost 20° for sources in the rear and elevated above the horizontal plane. Error in the apparent position data from the Wightman and Kistler study was about 18° and showed little dependence on source position. However, the stimulus spectrum in the Wightman and Kistler experiment was intentionally varied between trials, while it was stationary in the other studies. Use of a variable spectrum stimulus would not allow listeners to make optimal use of monaural spectral cues, since use of such cues requires a priori knowledge of the stimulus spectrum. In the Butler, Humanski, and Musicant study, stimulus positions and responses were constrained to one side of the head, and error was reported separately for the azimuth and elevation components of the judgments. Horizontal error was roughly constant at about 15° across the 180° range of azimuths studied. Vertical error varied from about 7° for azimuths near the front to about 14° for rear azimuths. In addition, vertical error was, on average, nearly twice as great at high source elevations than at low elevations. While comparison of these results with those from the other three studies is difficult, the overall pattern of results is qualitatively consistent across all four studies, suggesting that correspondence between the actual and apparent source position is best for sources in the front and worst for sources that are elevated and in the rear.

The acoustical determinants of apparent position are not equally effective in all frequency bands. This concept is, of course, the basic tenet of the duplex theory. The ITD cues are most likely to be effective in the low-frequency region, where they are both larger (Kuhn, 1977) and more salient (Blauert

1982; Trahiotis and Bernstein 1986). The ILD and monaural spectral cues are most likely to be effective in the high-frequency region, where they are considerably larger. Thus, if the spectrum of a broadband sound is restricted to low frequencies or high frequencies, the relative effectiveness of the various cues is revealed. For example, the experiment reported by Belendiuk and Butler (1978) suggested that, if a stimulus is low-passed at 1 kHz, ITDs must be present for accurate localization and, if the stimulus is high-passed above 4 kHz, ILDs are required. Accurate localization of sources on the vertical median plane is possible only if the stimulus includes frequencies above about 5 kHz (e.g., Roffler and Butler 1968). This is to be expected, since median plane localization depends on monaural spectral cues, and these cues, provided by pinna filtering, are most evident above 5 kHz (Shaw 1974, Middlebrooks, Makous, and Green 1989, Wightman and Kistler 1989b).

5.3 Sounds Heard Monaurally

One strategy many researchers have used to address questions about the effectiveness of monaural spectral cues is the presentation of stimuli to listeners with one ear plugged. The rationale is that monaural stimulus presentation eliminates the contribution of ITD and ILD cues, thus leaving only the monaural spectral cues to signal sound position. As noted in the introduction to this Section, there may be reasons to question the validity of this reasoning. It is well known that the apparent positions of such "monaural" stimuli are strongly lateralized to the side of the unplugged ear (Hebrank and Wright 1974a; Butler 1975; Blauert 1983; Oldfield and Parker 1986; Musicant and Butler 1980; Butler, Humanski, and Musicant 1990). Because the left-right position of a sound is governed entirely by interaural differences (Kistler and Wightman 1992), it does not seem reasonable to argue that interaural differences are "eliminated" by plugging one ear. It may be more appropriate to view the attenuation created by the plug as a large interaural difference cue that is fixed and independent of the monaural spectral cues. While the effect of the plug on ITDs is less certain, the attenuation caused by the plug clearly produces ILDs that are large even at low frequencies. Moreover, with one ear plugged, the interaural difference cues often conflict with the monaural spectral cues since, regardless of the actual source position, the extreme ILD always signals a position directly opposite the unoccluded ear.

Most of the research on monaural localization has focused not on the apparent position of sounds heard with one ear but on the accuracy with which listeners could identify source position. Given that sounds heard monaurally are lateralized to one side, it is not surprising that "accuracy" in judging horizontal location is said to be low for sources in the front or rear and high for sources at the side, near 90° azimuth (Oldfield and Parker 1986; Butler, Humanski, and Musicant 1990). As noted in Section 4, the high accuracy of monaural listeners judging the azimuth of sources near 90° is a result of the fact that the apparent azimuth of all monaural sounds is near 90°. The focus on localization accuracy obscures what is the most dramatic consequence of plugging one ear, namely, that the apparent positions of sounds heard monaurally are all opposite the unoccluded ear and that "localization" in the normal sense of the word is thus eliminated.

Error in judging the vertical component of source position is only moderately increased by plugging one ear, as long as the source spectrum is stationary from trial to trial (Hebrank and Wright 1974a; Oldfield and Parker 1986; Butler, Humanski, and Musicant 1990; Thus, the vertical position of sounds heard monaurally must be determined largely by monaural spectral cues. When the source spectrum is randomized from trial to trial, elevation errors increase (Hebrank and Wright 1974a; Wightman, Kistler, and Arruda 1991). In the Wightman, Kistler, and Arruda study with scrambled spectrum sounds, most listeners reported monaural sources to have elevations near 0°, so the vertical error was very high. Hebrank and Wright (1974a) describe a more modest increase in elevation error following randomization of the sound spectrum.

5.4 Virtual Sound Sources

A substantial number of studies directed at issues related to sound localization have involved the presentation of stimuli over headphones. The advantage of headphone stimulus presentation over free-field presentation is clear: headphones offer complete control of the acoustical stimulus and allow modification of stimulus parameters in ways that would not be feasible in free field presentation. However, headphone-presented sounds usually have apparent positions inside the head of the listener, and, for this reason, the first headphone experiments emphasized issues such as the detectability and discriminability of interaural differences. More recent lateralization experiments have measured the apparent positions of the internalized images, with an emphasis on the effect of changes in ITD and ILD (Yost 1981; Blauert 1982, 1983; Bernstein and Trahiotis 1985; Trahiotis and Bernstein 1986).

Typically, the ITDs and ILDs chosen for lateralization experiments are somewhat unnatural. For example, the ILD is usually constant across a wide frequency band, something that never occurs with sounds presented in free field because of pinna filtering effects. It is likely that the internalized character of the auditory images in lateralization experiments is a result primarily of the use of these unnatural stimulus parameters. Indeed, when headphone-presented sounds are given more natural free-field characteristics, especially those contributed by the pinnae, the sound image externalizes (Plenge 1974; Blauert 1983).

Wightman and Kistler (1989b, 1992) and Kistler and Wightman (1992) have reported psychophysical measurements of the apparent positions of headphone-presented sounds (called "virtual sources") to which the major naturally occurring directional cues such as ITDs, ILDs, and pinna effects had been added by digital filtering. The digital filters were derived from measurements of each individual subject's free field HRTF at a large number of spatial positions. Sound pressure waveform differences measured at each listener's eardrums between free-field sources and comparable virtual sources were minimal, suggesting that the synthesis faithfully captured all the potential acoustical determinants of apparent position (Wightman and Kistler 1989a). Psychophysical measurements confirmed the adequacy of the synthesis. Listeners' judgments of the apparent positions of virtual sources were indistinguishable from their judgments of the apparent positions of comparable free-field sources (Wightman and Kistler 1989b).

To the extent that virtual and real sources are equivalent, the virtual source technique can be used to address questions about free-field localization that are difficult or impossible to confront using real free-field sources. Since the stimulus waveform presented to each of a listener's ears can be readily controlled using headphones, the potential acoustical determinants of apparent sound position, such as ITDs, ILDs, and spectral cues, can be precisely and independently manipulated, something not generally feasible in free field. Wightman and Kistler (1992) and Kistler and Wightman (1992) have used the virtual source technique to study a number of questions about the relative salience of the various potential determinants of apparent sound position. In one study, wideband random-spectrum noise stimuli were synthesized such that the ITD cues conflicted with the ILD and spectral cues (Wightman and Kistler 1992). In other words, each stimulus contained natural directional cues derived from individualized HRTF measurements, but the ITD cues signaled a different source direction from the ILD and spectral cues. Listeners' judgments of the apparent directions of the conflicting cue stimuli consistently coincided with the direction signaled by the ITD cue. Even when the ILD and spectral cues indicated a direction 180° displaced from that indicated by the ITD cue, listeners appeared to base their judgments entirely on the ITD cue. Moreover, there was no evidence in the data that the conflicting cue stimuli evoked multiple or diffuse sound images. The variance of the judgments in the conflicting cue conditions was the same as in the normal virtual source conditions. With low frequencies removed from the stimuli by high-pass filtering, the dominance of the ITD cue was eliminated. While the necessary cutoff frequency is subject dependent, the data from all listeners show complete release from ITD dominance when frequencies below 5 kHz are filtered out.

The virtual source technique has also been used to show that the spectral cues provided by pinna filtering are important primarily for the perception of source elevation and for the resolution of front-back confusions (Kistler and Wightman 1992). This experiment involved stimuli synthesized with digital filters that were derived from a principal components analysis (PCA) of HRTFs. The procedure allowed systematic and controlled degradation of the spectral detail provided by pinna filtering. Listeners' judgments of the apparent positions of sources synthesized with the PCA technique revealed that even slight degradation of pinna effects caused increases in front-back confusions and noticeable distortions of elevation perception. The distortions in elevation perception that accompanied the degradation of pinna filtering have been reported before in free-field experiments in which the folds of listeners' pinnae were filled with putty or the pinnae were deformed (Roffler and Butler 1968; Gardner and Gardner 1973; Butler 1975). Remarkably, the left-right component of apparent image position (azimuth irrespective of front-back position) was undisturbed, even by complete removal of all peaks and valleys in the HRTFs used to synthesize the stimuli. Since removal of HRTF peaks and valleys had no effect on ITDs and did not change overall ILDs, it can be concluded that the left-right component of apparent position is determined primarily by interaural differences.

In certain laboratory conditions, localization of virtual sources is indistinguishable from localization of real sources. However, there can be no denying the fact that, in everyday localization, there are important determinants of apparent sound position, such as those provided by head movement, visual referents, etc., that are missing from current implementations of the virtual source technique. The advantages of the virtual source technique for experimental use are nevertheless quite substantial, so it is likely they will see much wider use in the future.

5.5 Individual Differences

The acoustical determinants of apparent sound position can be expected to vary considerably with individual anatomical features such as head size, pinna breadth and depth, etc., since those features have dimensions roughly comparable to the wavelengths of sounds in the range of human hearing. In adults, variability in upper body dimensions is substantial. Burkhard and Sachs (1975), for example, reported measurements from 12 males and 12 females in which the mean concha length is 2.63 cm, mean breadth is 1.8 cm, with standard deviations in both dimensions of about 0.23 cm. This implies that in the adult population concha lengths could vary between about 2.1 and 3.1 cm and breadths from about 1.3 to 2.3 cm ($\pm 2\sigma$). Since the resonances of the concha play such a major role in outer ear acoustics (Shaw and Teranishi 1968), size variations of 20% to 30% about the mean are likely to be significant.

There is ample evidence that individual differences in anatomical dimensions translate to individual differences in the potential acoustical cues for sound localization. For example, Middlebrooks and Green (1990) showed that head radius varied from about 6.1 to 7.3 cm in their sample of six adults and, as a result, the maximum ITD varied from about 645 to 750 μs. Thus, an ITD of 645 μs would signal a source at 90° azimuth for the subject with the smaller head but would indicate a source at only about 75° for the other subject. Shaw and Teranishi (1968) reported that, for a frontal sound source, the frequency coordinates of the peaks and notches in the pinna frequency response varied by as much as 4 kHz in a group of six adults. A 4-kHz shift in the frequency of a spectral notch normally indicates an elevation change of over 60° for a frontal source (Shaw and Teranishi 1968).

Given the large individual differences in acoustical cues that are an inevitable consequence of anatomical variation, it is somewhat surprising that there have been so few reports of large individual differences in sound localization behavior. To the contrary, most studies of the apparent positions of broadband sounds do not report variation from subject to subject in the general pattern of results, suggesting that it was small. In the recent Wightman and Kistler (1989b) study, however, in which individual data are shown, the individual differences are clear. The data from one subject in particular suggested that apparent elevation was perceived very differently by this subject. Analysis of that subject's HRTFs suggested that the degraded elevation perception might be a result of a substantial lack of pinna directionality in the vertical dimension.

Given that the apparent positions of broadband sounds usually match the actual positions quite well, in spite of substantial variability in the acoustical cues, the logical conclusion is that listeners learn the associations between sound positions and the corresponding acoustical cues as provided by their own individual outer ear structures (as suggested by Thompson 1882 and quoted in Section 3.2). The best empirical confirmation of this hypothesis comes from experiments in which listeners judge the apparent position of virtual sounds that have been synthesized using either their own HRTF measurements or those from someone else (Wenzel et al. 1988; Wenzel, Wightman, and Kistler 1991; Wenzel et al. 1993). In every case, listening "through someone else's ears" resulted in changes in the apparent positions of sounds so as to decrease the correspondence between apparent position and actual position. Moreover, only the elevation components of the judgments and the frequency of front-back confusions appeared to be affected, implying that the slight variations in ITD as a result of head size variations were of lesser consequence.

6. Conclusions

Modern research reveals that the apparent position of an auditory object is determined by a rich set of cues. The set includes not only the long-appreciated interaural differences and spectral cues but also visual, contextual, and cognitive cues. The relative salience of any one cue, and the influence of interactions among the cues, depend on a number of factors, including the sound's frequency content and familiarity, the nature of other sounds in the environment, and the presence or absence of a corresponding visual referent. It seems clear that the primary cues are the interaural differences, and that, in most situations, the left-right component of apparent position (arguably the most "important" component of object position for human listeners) is determined by these interaural differences. The other cues are secondary and important primarily for resolving ambiguities such as front-back confusions and for determining the apparent elevation of an object. There are few, if any, data that contradict the duplex theory which argues for the salience of ITDs at low frequencies and ILDs at high frequencies. However, recent experiments suggest that the theory needs to be elaborated to incorporate the finding that, under certain conditions, low-frequency ITD cues can actually override high-frequency ILD cues if the two conflict.

Future research will surely clarify the complex interactions among the acoustical cues to sound position and reveal the conditions under which the nonacoustical cues are important. In addition, perfection of the virtual source techniques will almost certainly stimulate systematic study of a number of important issues that have yet to be resolved because of technical difficulties. Among these are the factors that determine the apparent distance of an auditory object, the role of reverberation and echoes, and the perception of moving objects. There can be little doubt that, in the years ahead, some of the most perplexing aspects of human sound localization will yield to the ever more powerful research tools of the modern hearing scientist.

Acknowledgements. Preparation of this chapter was supported in part by a grant from the NIH-NIDCD and by a Cooperative Research Agreement (NCCA-542) with NASA-Ames Research Center. We are grateful to Ms. Marianne Arruda for her long-term assistance with all aspects of our research, to Dr. Mark Stellmack and Mr. Ewan Macpherson for their careful editing, and to Dr. Robert Lutfi for his extensive and thoughtful comments on earlier versions of this chapter.

References

Angell JR, Fite W (1901) The monaural localization of sound. Psych Rev 8:225–246.
Batteau DW (1967) The role of the pinna in human localization. Proc R Soc London Ser B 168:158–180.
Belendiuk K, Butler RA (1975) Monaural localization of low-pass noise bands in the horizontal plane. J Acoust Soc Am 58:701–705.
Belendiuk K, Butler RA (1978) Directional hearing under progressive impoverishment of binaural cues. Sensory Processes 2:58–70.
Bernstein LR, Trahiotis C (1985) Lateralization of sinusoidally amplitude-modulated tones: Effects of spectral locus and temporal variation. J Acoust Soc Am 78:514–523.
Blauert J (1969) Sound localization in the median plane. Acustica 22:205–213.
Blauert J.(1971) Localization and the law of the first wavefront in the median plane. J Acoust Soc Am 50:466–470.
Blauert J (1982) Binaural localization: Multiple images and applications in Room- and electroacoustics. In: Gatehouse WR (ed) Localization of Sound: Theory and Applications. Groton, CN: Amphora Press, pp. 65–84.

Blauert J (1983) Spatial Hearing. Cambridge, MA: MIT Press.
Bloom PJ (1977) Creating source elevation illusions by spectral manipulation. J Audio Eng Soc 25:560–565.
Burkhard MD, Sachs RM (1975) Anthropometric manikin for acoustic research. J Acoust Soc Am 58:214–222.
Butler RA (1969) On the relative usefulness of monaural and binaural cues in locating sound in space. Psychon Sci 17:245–246.
Butler RA (1975) The influence of the external and middle ear on auditory discriminations. In: Keidel WD, Neff WD (eds) Handbook of Sensory Physiology. Berlin: Springer-Verlag, pp. 247–260.
Butler RA (1986) The bandwidth effect on monaural and binaural localization. Hear Res 21:67–73.
Butler RA, Belendiuk K (1977) Spectral cues utilized in the localization of sound in the median sagittal plane. J Acoust Soc Am 61:1264–1269.
Butler RA, Flannery R (1980) The spatial attributes of stimulus frequency and their role in monaural localization of sound in the horizontal plane. Percept Psychophys 28:449–457.
Butler RA, Humanski RA (1992) Localization of sound in the vertical plane with and without high-frequency spectral cues. Percept Psychophys 51:182–186.
Butler RA, Planert N (1976) The influence of stimulus bandwidth on localization of sound in space. Percept Psychophys 19:103–108.
Butler RA, Humanski RA, Musicant AD (1990) Binaural and monaural localization of sound in two-dimensional space. Perception 19:241–256.
Colburn HS (1973) Theory of binaural interaction based on auditory-nerve data. I. General strategy and preliminary results on interaural discrimination. J Acoust Soc Am 54:1458–1470.
Coleman PD (1962) Failure to localize the source distance of an unfamiliar sound. J Acoust Soc Am 34:345–346.
Durlach NI, Colburn HS (1978) Binaural Phenomena. In: Carterette EC, Friedman MP (eds) Handbook of Perception, Volume IV—Hearing. New York: Academic Press, pp. 365–466.
Feddersen WE, Sandel TT, Teas DC, Jeffress LA (1957) Localization of high-frequency tones. J Acoust Soc Am 29:988–991.
Firestone FA (1930) The phase difference and amplitude ratio at the ears due to a source of pure tones. J Acoust Soc Am 2:260–270.
Fisher HG, Freedman SJ (1968) The role of the pinna in auditory localization. J Aud Res 8:15–26.
Gardner MB (1969) Distance estimation of 0° or apparent 0°-oriented speech signals in anechoic space. J Acoust Soc Am 45:47–53.
Gardner MB, Gardner RS (1973) Problem of localization in the median plane: Effect of pinnae cavity occlusion. J Acoust Soc Am 53:400–408.
Grantham DW (1986) Detection and discrimination of simulated motion of auditory targets in the horizontal plane. J Acoust Soc Am 79:1939–1949.
Green DM (1971) Temporal auditory acuity. Psych Rev 78:540–551.
Green DM (1976) An Introduction to Hearing. New York: John Wiley and Sons.
Hartmann WM (1983) Localization of sound in rooms. J Acoust Soc Am 74:1380–1391.
Hartmann WM, Rakerd B (1989a) On the minimum audible angle—A decision theory approach. J Acoust Soc Am 85:2031–2041.

Hartmann WM, Rakerd B (1989b) Localization of sound in rooms. IV: The Franssen effect. J Acoust Soc Am 86:1366–1373.

Hebrank JH, Wright D (1974a) Are two ears necessary for localization of sound sources on the median plane? J Acoust Soc Am 56:935–938.

Hebrank JH, Wright D (1974b) Spectral cues used in the localization of sound sources on the median plane. J Acoust Soc Am 56:1829–1834.

Jackson CV (1953) Visual factors in auditory localization. Quart J Exp Psych V:52–65.

Joris PX, Yin TCT (1992) Responses to Amplitude-modulated tones in the auditory nerve of the cat. J Acoust Soc Am 91:215–232.

Kistler DJ, Wightman FL (1992) A model of head-related transfer functions based on principal components analysis and minimum-phase reconstruction. J Acoust Soc Am 91:1637–1647.

Klumpp R, Eady H (1956) Some measurements of interaural time differences thresholds. J Acoust Soc Am 28:859–864.

Kuhn GF (1977) Model for the interaural time differences in the azimuthal plane. J Acoust Soc Am 62:157–167.

Kuhn GF (1987) Physical Acoustics and Measurements Pertaining to Directional Hearing. In: Yost WA, Gourevitch G (eds) Directional Hearing. New York: Springer-Verlag, pp. 3–25.

Lindemann W (1986) Extension of a binaural cross-correlation model by contralateral inhibition. I. Simulation of lateralization for stationary signals. J Acoust Soc Am 80:1608–1622.

Makous JC, Middlebrooks JC (1990) Two-dimensional sound localization by human listeners. J Acoust Soc Am 87:2188–2200.

Mehrgardt S, Mellert V (1977) Transformation characteristics of the external human ear. J Acoust Soc Am 61:1567–1576.

Mellert V, Siebrasse KF, Mehrgardt S (1974) Determination of the transfer function of the external ear by an impulse response measurement. J Acoust Soc Am 56:1913–1915.

Middlebrooks JC (1992) Narrowband sound localization related to external ear acoustics. J Acoust Soc Am 92:2607–2624.

Middlebrooks JC, Green DM (1990) Directional dependence of interaural envelope delays. J Acoust Soc Am 87:2149–2162.

Middlebrooks JC, Green DM (1991) Sound localization by human listeners. Ann Rev Psychol, 42:135–159.

Middlebrooks JC, Green DM (1992) Observations on a principal components analysis of head-related transfer functions. J Acoust Soc Am 92:597–599.

Middlebrooks JC, Makous JC, Green DM (1989) Directional sensitivity of sound-pressure levels in the human ear canal. J Acoust Soc Am 86:89–108.

Mills AW (1958) On the minimum audible angle. J Acoust Soc Am 30:237–246.

Mills AW (1960) Lateralization of high-frequency tones. J Acoust Soc Am 32:132–134.

Mills AW (1972) Auditory localization. In: Tobias JV (ed) Foundations of Modern Auditory Theory. New York, London: Academic Press, pp. 303–348.

Musicant AD, Butler RA (1980) Monaural localization: An analysis of practice effects. Percept Psychophys 28:236–240.

Musicant AD, Butler RA (1985) Influence of monaural spectral cues on binaural localization. J Acoust Soc Am 77:202–208.

Neti C, Young ED, Schneider MH (1992) Neural network models of sound localization based on directional filtering by the pinna. J Acoust Soc Am 92:3140–3156.

Oldfield SR, Parker SPA (1984a) Acuity of sound localization: A topography of auditory space. I. Normal hearing conditions. Perception 13:581–600.

Oldfield SR, Parker SPA (1984b) Acuity of sound localization: A topography of auditory space. II. Pinna cues absent. Perception 13:601–617.

Oldfield SR, Parker SPA (1986) Acuity of sound localization: A topography of auditor space. III. Monaural hearing conditions. Perception 15:67–81.

Parducci A (1963) Range-frequency compromise in judgment. Psychological Monographs, 77 No. 2:1–60.

Perrott DR, Musicant AD (1977) Minimum auditory movement angle: Binaural localization of moving sound sources. J Acoust Soc Am 62:1463–1466.

Pick HL, Warren DH, Hay JC (1969) Sensory conflict in judgments of spatial direction. Percept Psychophys 6:203–205.

Plenge G (1974) On the differences between localization and lateralization. J Acoust Soc Am 56:944–951.

Pollack I, Rose M (1967) Effects of head movements on the localization of sounds in the equatorial plane. Percept Psychophys 2:591–596.

Pratt CC (1930) The spatial character of high and low tones. J Exp Psych 13:278–285.

Rakerd B, Hartmann, WM (1985) Localization of sound in rooms. II: The effects of a single reflecting surface. J Acoust Soc Am 78:524–533.

Rakerd B, Hartmann WM (1986) Localization of sound in rooms. III: Onset and duration effects. J Acoust Soc Am 80:1695–1706.

Roffler SK, Butler RA (1968) Factors that influence the localization of sound in the vertical plane. J Acoust Soc Am 43:1255–1259.

Sandel TT, Teas DC, Feddersen WE, Jeffress LA (1955) Localization of sound from single and paired sources. J Acoust Soc Am 27:842–852.

Searle CL, Braida LD, Davis MF, Colburn HS (1976) Model for auditory localization. J Acoust Soc Am 60:1164–1175.

Shaw EAG (1965) Ear canal pressure generated by a free sound field. J Acoust Soc Am 39:465–470.

Shaw EAG (1974) Transformation of sound pressure level from the free field to the eardrum in the horizontal plane. J Acoust Soc Am 56:1848–1861.

Shaw EAG, Teranishi R (1968) Sound pressure generated in an external ear replica and real human ears by a nearby sound source. J Acoust Soc Am 44:240–249.

Shelton BR, Searle CL (1980) The influence of vision on the absolute identification of sound-source position. Percept Psychophys 28:589–596.

Stern RM, Colburn HS (1978) Theory of binaural interaction based on auditory-nerve data. IV. A model for subjective lateral position. J Acoust Soc Am 64:127–140.

Stevens SS, Newman EB (1936) The localization of actual sources of sound. Am J Psych 48:297–306.

Strutt JW, Lord Rayleigh (1907) On our perception of sound direction. Philos Mag 13:214–232.

Thompson SP (1882) On the function of the two ears in the perception of space. Philos Mag 13:406–416.

Thurlow WR, Runge PS (1967) Effect of induced head movements on localization of direction of sounds. J Acoust Soc Am 42:480–488.

Trahiotis C, Bernstein LR (1986) Lateralization of bands of noise and sinusoidally amplitude-modulated tones: Effects of spectral locus and bandwidth. J Acoust Soc Am 79:1950–1957.

Trimble OT (1934) Localization of sound in the anterior-posterior and vertical dimensions of auditory space. Br J Psych 24:320–334.

Wallach H (1939) On sound localization. J Acoust Soc Am 10:270–274.

Wallach H (1940) The role of head movements and vestibular and visual cues in sound localization. J Exp Psych 27:339–368.

Warren DH (1970) Intermodality interactions in spatial localization. In: Reitman W (ed) Cognitive Psychology. New York: Academic Press, pp. 114–133.

Watkins AJ (1978) Psychoacoustical aspects of synthesized vertical locale cues. J Acoust Soc Am 63:1152–1165.

Wenzel EM, Wightman FL, Kistler DJ, Foster SH (1988) Acoustic origins of individual differences in sound localization behavior. J Acoust Soc Am 84:S79.

Wenzel EM, Wightman FL, Kistler DJ (1991) Localization with non-individualized virtual acoustic display cues. In: Proceedings of CHI '91, ACM Conference on Computer-Human Interaction. New York: ACM Press, pp. 351–359.

Wenzel EM, Arruda M, Kistler DJ, Wightman FL (1993) Localization using non-individualized head-related transfer functions. J Acoust Soc Am 94:111–123.

Wiener FM (1947) On the diffraction of a progressive sound wave by the human head. J Acoust Soc Am 19:143–155.

Wiener FM, Ross DA (1946) The pressure distribution in the auditory canal in a progressive sound field. J Acoust Soc Am 18:401–408.

Wightman FL, Kistler DJ (1989a) Headphone simulation of free-field listening. I: Stimulus synthesis. J Acoust Soc Am 85:858–867.

Wightman FL, Kistler DJ (1989b) Headphone simulation of free-field listening. II: Psychophysical validation. J Acoust Soc Am 85:868–878.

Wightman FL, Kistler DJ (1992) The dominant role of low-frequency interaural time differences in sound localization. J Acoust Soc Am 91:1648–1661.

Wightman FL, Kistler DJ, Arruda M (1991) Monaural localization, revisited. J Acoust Soc Am 89:1995.

Woodworth RS (1938) Experimental Psychology. New York: Holt,

Yin TCT, Chan JCK (1988) Neural mechanisms underlying interaural time sensitivity to tones and noise. In: Edelman GM, Gall WE, Cowan WM (eds) Auditory Function: Neurobiological Bases of Hearing. New York: John Wiley and Sons, pp. 385–430.

Yost WA (1981) Lateral position of sinusoids presented with interaural intensive and temporal differences. J Acoust Soc Am 70:397–409.

Yost WA, Gourevitch G (1987) Directional Hearing. New York; Springer-Verlag.

Yost WA, Hafter ER (1987) Lateralization. In: Yost WA, Gourevitch G (eds) Directional Hearing. New York: Springer-Verlag, pp. 49–84.

6
Auditory Perception

WILLIAM A. YOST AND STANLEY SHEFT

1. Sound Source Determination and Auditory Perception

Most of the material covered in the previous four chapters has described the sensitivity of human listeners to the basic physical variables of sound responsible for hearing. The data and theories resulting from this work have defined both the form of the auditory code for sound and some of the mechanisms responsible for generating that code. However, if we are to determine the sources of the sounds in our world, we need to know more than just the sensitivity of the auditory system to the frequency, intensity, and time/phase characteristics of sound. In the past decade, there has been an emerging realization that processing the complex sounds that make up our everyday lives is not always predictable from what is known about a listener's sensitivity to the basic properties of sound and that, most likely, there are unifying themes that can be applied to the perception of all sounds, including speech and music (Yost and Watson 1987).

One such unifying theme has resurfaced. It was described by Yost (Chapter 1) as sound source determination, but Yost (1991), in a review, labeled the theme auditory image analysis, Bregman (1990), in his book, referred to it as auditory scene analysis, Hartmann (1988) has called it the perception of auditory entities, and Moore (1989) referred to the topic as auditory object perception. All of these writers expressed essentially the same viewpoint: hearing allows an organism to determine the sources of sound in its environment. The sounds from individual sources do not arrive at the auditory periphery as separate acoustic events but rather as one complex sound field (see Yost, Chapter 1, Fig. 1). Furthermore, it is known that the auditory periphery first resolves the vibrations of that complex sound field into its spectral components represented by a spectro-temporal response pattern. This response pattern has to be further processed by the auditory system for the information available in the spectro-temporal code to reveal the source of the sounds. To many, this processing constitutes auditory perception. [See Bregman (1990) for an explicit statement of this position and Handel (1989) for an implicit discussion.] Since we do not have direct auditory access to the sound sources, the sources are deduced from our perceptions. Bregman (1990) uses a visual analogy to describe our perception as an auditory scene of images, where each image is a sound source or potential sound source. It is easy to see how such a view of auditory perception can quickly venture into areas of cognition and phenomenology. It is also tempting to use analogies to vision, especially to principles from the Gestalt school of visual perception, to help explain auditory scene analysis.

Though the analogy to vision does help clarify the issue of auditory scene analysis and sound source determination, consideration of the sensory coding at the peripheral level suggests different processing schemes for the two senses. In vision, a single object in space tends to stimulate only a limited region of the sensory macula, with much of response of the lower nervous system being analytic in its sensitivity to various features of the object. The challenge for visual scene analysis then becomes one of synthesis, incorporating features to perceive the source. In audition, a single source can often stimulate a wide region of the sensory macula, with multiple sources simultaneously stimulating the same peripheral frequency channels. Though specific units in the central auditory nervous system (CANS) exhibit tuned sensitivity to specific acoustic features, sound source determination requires analytic processing which segregates spectral components according to source.

This chapter will neither wander beyond the focus of auditory perception nor will it rely on visual analogies. We will tend to avoid the use of images, objects, or entities to describe this view of auditory perception but will continue to use sound source determination since it refers to the initial goal of the listener. We will avoid most of the issues of cognition and phenomenology by focusing on what information might be used as the basis of sound source determination.

Let's consider the following example that has been referred to as the two-vowel problem. If two people were each to utter a different vowel at the same time, most listeners would have little difficulty knowing that the two different vowels were spoken by two different people. In developing both speech synthesis and speech analysis systems, two simulated vowels may not be recognized by listeners when they are mixed or natural vowels may not be computer-recognized when they are mixed. McAdams (1984) has provided an interesting example: two vowels are synthesized, each with a perfectly periodic sound source simulating a perfectly periodic vocal cord function. When the proper attempts to model the vocal tract properties for each vowel are made, each vowel when presented alone is recognized as such and is judged to be a good exemplar of speech. However, when the two vowels are mixed, neither vowel can be heard and often the mixed sound does not even sound like speech. Thus, under these conditions, listeners can not perceive the two vowels as they would in normal listening situations. Obviously, the simulation is lacking some key ingredients. If the vocal cord sound source for one vowel is made aperiodic (i.e., the time intervals between the simulated glottal pulses are randomly jittered to simulate naturally produced speech) rather than periodic, the two vowels can then be heard again in the mixture.

The two-vowel example will help demonstrate the key elements in sound source determination as it applies to auditory perception. In Section 7, we will discuss this stimulus condition as an example of the use of coherent temporal modulation aiding sound source determination. The sound sources in this example are the vocal cord and tract variables characterizing the two vowel sounds (since the two vowels were synthesized as if the vowels were produced by two different people, the sources might be described as the two people). The mixture of the two sounds constitutes an example of a complex sound field. Sound source determination is the ability to correctly identify the two vowels. In this example, the determination of the sources is achieved by identifying the two vowels. However, it is not necessary for a listener to identify a source in order to determine that a source exists. That is, a novel sound may not be identifiable (i.e., have a label assigned to it), but it could still be perceived as a sound source. Identification of the sound (labeling it based on one's past experience) is neither necessary nor sufficient for sound source determination. In this example, one might argue that the ability of the listener to determine the two sources is based on his/her past experience with speech and is, thus, a type of cognitive or learning (or top-down) process, one that has no major hearing or psychoacoustic (or bottom-up) component. However, the listener can only determine the sources when the synthesized source for one vowel was made aperiodic. The auditory system appears to use the fact that all of the frequency components associated with one of the vowels in the two-vowel mixture can be fused based on the common pattern of aperiodic glottal pulse production. Even if there are aspects of sound source determination that are based on the past experience of the listener, the acoustic information in the sound emanating from each source must have certain characteristics that allow for source determination. Thus, as the first step in understanding sound source determination, a description of these characteristics appears crucial.

The study of sound source determination has probably not been a major topic for the hearing sciences for a number of historical reasons (see Boring 1942). The modern study of sensory systems has its roots in the middle of the nineteenth century when philosophers and scientists wrestled with dualism. A topic of debate was sensation versus perception, with perception tied to the objects of the world and sensation to the attributes of the environmental stimulus. Sound, as the stimulus for hearing, was not considered an object. For instance, sound was not believed to be truly localized in space, because only physical objects had extent and thus the dimensions of space. The fact that sound sources could be localized was explained by arguing that the perception of auditory space was derived through its association with the other senses (vision and touch) and with consciousness. By the time actual sound localization experiments were reported by Lord Rayleigh (1877/1945) and others (again see Boring 1942), the interest in hearing had switched to the realization that the auditory system transformed sound into a pattern of pitch impressions. As Helmholtz (1885/1954) and others proposed various "theories of hearing" to explain this transformation, physiologists and anatomists were beginning to study the cochlea directly. The role of the auditory periphery in determining the frequency content of sound took over as the compelling question for audition.

In order to study sound source determination, an investigation of the physical variables, in addition to just the frequency content of sound is needed. The following sections of this chapter review some of what is known about six possible physical variables that might be used for sound source determination: Spectral Separation (Section 2.); Spectral Profiles (Section 3.); Harmonicity (Section 4.); Spatial Separation (Section 5.); Temporal Separation (Section 6.); and Coherent Temporal Modulation (Section 7.). The chapter then covers some of the literature that has dealt either directly or indirectly with issues pertaining to Fusion and Segregation (see Yost, Chapter 1, Table 1).

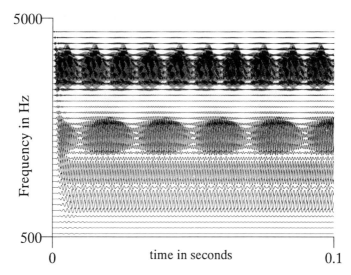

FIGURE 6.1. The response of the filter bank stage of the Patterson and Holdsworth (Patterson et al. 1992) model to a six-tone complex. The six components of the complex are the first, second, and fourth harmonics of an 880-Hz fundamental and the second, third, and sixth harmonics of a 550-Hz fundamental. The outputs of these filters provide an estimate of the spectral information available to the auditory periphery. If three components came from one source and three from another, it would be difficult for the auditory system to segregate the information based on spectrum alone. The close spacing between high-frequency components does not allow for complete resolution by peripheral filtering, as evidenced by the slow modulation of the high center-frequency filter outputs.

2. Spectral Separation

The auditory system's remarkable ability to determine the spectral content of sound provides one form of information that might be used for sound source determination. Moore (Chapter 3) discussed a great deal of the psychophysical evidence that describes the frequency selectivity of the auditory system. Thus, in a simple case when two or more sounds contain different spectral components, the auditory system resolves those components. If the resolvable components from one source are clearly separable from those of another source, this separation might be the basis for sound source determination.

The sounds from most sources would not conform to this simple example. As Figure 6.1 shows, sounds that have high-frequency components that are close together will not be resolved by the auditory periphery because the bandwidths of the high-frequency auditory filters are too broad. When the frequency components associated with different sources are close together or overlap, as shown in the examples of Figure 6.1, it is not apparent how frequency selectivity can be the sole basis for sound source determination.

Thus, frequency selectivity and spectral separability alone do not appear as appealing variables that directly control sound source determination, except in some particular circumstances. However, the fact that sounds are resolved into their spectral components is crucial for many of the other variables discussed in the rest of this chapter to operate.

3. Spectral Profiles

The spectra of sound sources are characterized by their pattern of intensity variation as a function of frequency. These spectral patterns are often relatively invariant across changes in the level of source output. Thus, source determination requires an ability to process the spectral pattern or profile of the source output independent of overall level. The material reviewed by Green (Chapter 2) described some of the recent research on profile analysis performed primarily by Green and his students and colleagues (1988). In one set of conditions of profile-analysis experiments, a spectral profile is generated with one tonal component of a multitone complex having a greater level than the other components. The listener's task is to detect this level increment when the overall level of the complex is randomly varied over a large range. The results show that the listener is very sensitive to the level increment in some conditions. The ability to detect the increment is poor when there are few components (a low density of spectral components in the complex), improves when the number of components is increased, and then becomes poor again as the density of the components increases further. Figure 6.2 shows that the notion of detecting the spectral profile is somewhat consistent with what probably occurs within the peripheral auditory system. When the components are few and widely spaced, the spectral peak is not as easily

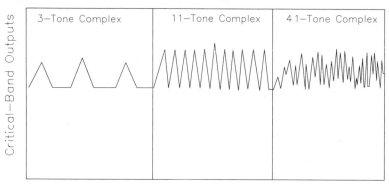

FIGURE 6.2. A schematic diagram of the excitation patterns generated by three different spectral profiles used in profile experiments. In all three, the middle tone is more intense. In the first pattern, there is a 3-tone profile, in the second pattern an 11-tone profile, and in the third pattern a 41-tone profile. The intensity increment is easiest to detect in the second spectral pattern. The profile is less clear in the first pattern because there are too few components spaced too far apart. It is difficult to detect the intensity increment in the pattern for the 41-tone profile because many spectral components fall within a single critical band resulting in masking.

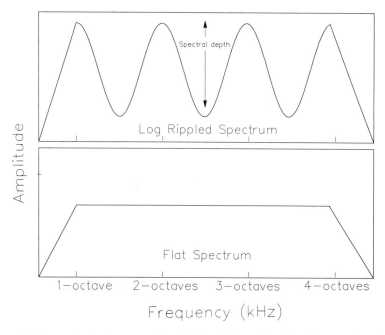

FIGURE 6.3. An example of a spectral profile in which amplitude varies as a sinusoidal function of the logarithm of frequency. The task is to discriminate between the rippled (top) and flat (bottom) profiles, with the depth of the spectral ripple adjusted to estimate threshold. Listeners are very sensitive to such ripples, even when the overall level of both profiles is randomly varied over a large range (e.g., 40 dB).

discernable as when the density of the components is higher. When the density becomes too great, the components overlap within a single auditory channel or critical band and mask the increment.

More complex spectral profiles have also been studied. Figure 6.3 indicates a test stimulus that consists of a spectral profile that varies sinusoidally (over log frequency) with a spectral period of variation of approximately one cycle per octave. The peak-to-valley ratio of this sinusoidal spectral profile can be adjusted until listeners are at threshold in their ability to discriminate the rippled stimulus from a stimulus in which all of the components are presented at the same level. Even when the overall level of the stimulus is randomly varied over a 40-dB range, listeners can still detect a peak-to-valley ratio of 2 dB or less in the rippled spectrum (Bernstein and Green 1988). Listeners are, in fact, better at detecting this ripple than one would expect based on how well they detect an increment at a single frequency. Although a number of different mechanisms may underlie the ability to process these complex spectra, the fact remains that listeners are quite sensitive at detecting changes in the spectral pattern of a complex sound even when the overall level of the sound undergoes large level variations. This sensitivity would seem to be a prerequisite to accurately determine the sources of sound.

3.1 Timbre

The general description of the perception of these stimuli with different complex spectra is that they differ in timbre (or, in some cases, in pitch; see Section 4 on Harmonicity). Timbre has always been a difficult concept to pin down. In general, timbre is described as the quality of a sound that allows one to discriminate one sound from another sound when both have the same pitch, loudness, and duration. Thus, a violin and a viola might play the same pitched note at the same loudness for the same duration, and we know that one instrument is the violin and one the viola because the two instruments produce different timbres. Scales of timbre range from simple descriptions to multidimensional scales, but, in general, sounds with different spectra often have different timbres (see Hirsh 1988). Thus, spectral profile is the stimulus variable most responsible for timbre. Clearly, our ability to process differences in timbre is crucial to distinguish one sound source from another (e.g., the violin and viola example given above).

Although this volume is not meant to review the speech perception literature, the relationship among spectral profiles, timbre, and speech (especially vowels) must be mentioned. Two vowels uttered by the same speaker at the same loudness (and most likely with essentially the same duration) are easily discriminated and thus, according to the definition given above, the two vowels could be said to differ in timbre. The discriminability of these two vowels shows little change if their overall levels are randomly varied over a large range. It is clear that the auditory system is processing the spectral profile of the vowel in essentially the same way that Green (1988) and his colleagues have described the processing of their "profile stimuli." A great deal of research has been directed at understanding what aspects of vowel spectra allow for vowel recognition (see Miller 1984, for review). Characterizing spectral profile analysis as one of the sound determination tasks then provides a way of integrating a large body of work in both auditory and speech perception (e.g., vowel perception, timbre perception, and complex spectral processing).

4. Harmonicity

A major aspect of most vibratory objects is that the vibration has a fundamental frequency with many higher harmonics. Green (Chapter 2) explained the data and theories associated with our perceptions of harmonic sounds; that is, a sound consisting of the harmonics of a fundamental are usually perceived as having a single pitch equal to that of the fundamental even if the fundamental is not in the sound's spectrum. This is a strong example of the synthesis or fusion of spectral components into an image or entity, in this case described as a pitch, where the perceived pitch could indicate the existence of a source such as the string of a violin or the voice of a speaker.

A great deal is known about the stimulus conditions which yield the perception of these complex pitches (see Moore, Chapter 3), often referred to as virtual pitches, to distinguish them from the pitches of sounds that are derived from the direct correspondence of the perceived pitch to the frequency location of energy in the spectrum (spectral pitch). These data demonstrate that there is a wide variety of stimulus conditions in which a complex sound consisting of many different spectral components is perceived as having a unitary pitch. The theories of complex pitch formation describe possible processes by which the auditory system fuses the spectral components into a pitch, and these theories will be described in more detail in Section 4.1.

Although the formation of virtual pitch is an excellent example of fusion, these pitches do not always allow for segregation of sound sources. That is, a complex sound field consisting of two harmonic series, as depicted in Figure 6.4, is not always perceived as having two pitches. Nor is a sound consisting of a harmonic series intermixed with spectral components that are not harmonically related always perceived as two sources, one having a pitch and one without a pitch (e.g., a noise). The process of forming virtual pitch is generally more synthetic than analytic. For the example shown in Figure 6.4, the system would most likely attempt to group the entire spectrum into one pitch (e.g., for Fig. 6.4, a listener might assign a pitch of 100 Hz to the stimulus because the first three harmonics of 100 Hz are present in the total complex) rather than analyze it into constituent parts (i.e., a complex sound field with two pitches that might correspond to two sources).

There are, however, situations in which a complex sound consisting of two different harmonic series is perceived as having two pitches. Beerends and

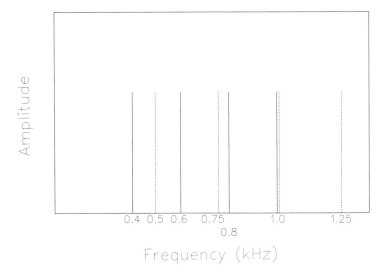

FIGURE 6.4. The spectrum of a sound that might be generated by two sound sources, one made up of the harmonics of 250 Hz (represented by the dashed spectral lines) and one made up of the harmonics of 200 Hz (represented by the solid spectral lines). Listeners are more likely to either detect no pitch or one closer to 100 Hz (since the first three components are harmonics of 100 Hz) than two pitches of 200 and 250 Hz.

Houtsma (1986, 1989) have measured the ability to identify the virtual pitch of two concurrent two-tone complexes. Results were used to estimate the internal variance in the coding of virtual pitch. Compared to estimates obtained in a single pitch identification task, variance estimates were greater in coding two simultaneous pitches. If the harmonics of one fundamental are presented to one ear and those of another fundamental to the other ear (i.e., the two harmonic sounds are spatially separated), then two virtual pitches can be heard under some conditions (Deutsch 1980). In other dichotic conditions, the auditory system appears to combine the information from the two ears to produce a "central pitch" (see Moore, Chapter 3). In these cases, incomplete harmonic spectral information is presented to each ear, and only by combining the information from the two ears could one account for the reported pitch (Houtsma and Goldstein 1972). Rasch (1978) and Darwin and Ciocca (1992) have shown that delaying the onset of one of two concurrent harmonic series enhances the ability to hear the two virtual pitches. Thus, if some other means of segregating the two harmonic series is used (e.g., spatial separation or onset asynchrony), then two pitches are, under some conditions, perceived when the two sets of harmonics are presented simultaneously.

When one of the tones of a harmonic series is varied in frequency by approximately 8% (made inharmonic), two pitches can be heard, one associated with the inharmonic tone and one associated with the fundamental of the harmonic series (Moore, Peters, and Glasberg 1985; Hartmann, McAdams, and Smith 1990; Moore and Glasberg 1990). A frequency shift of up to 3% will maintain the single virtual pitch of the fundamental. Hartmann (1988) showed that listeners can match the pitch of a harmonic mistuned by 4% to a pure tone. The matches are quite good for the lower harmonics of a 200-Hz fundamental and become less accurate when the mistuned harmonic is one of the higher harmonics. These and other data support the notion that a harmonic source will be heard as a single pitch unless a harmonic is mistuned by more than 3% to 8%. With greater mistuning, the complex will appear as two sources: one having the pitch of the fundamental and one the pitch of the mistuned harmonic.

The data for both the ability to hear out a mistuned harmonic of a harmonic series of tones and the ability to hear two virtual pitches when two harmonic series are mixed together has been explained in terms of a "harmonic sieve" model. The basis of this model is shown in Figure 6.5. If the frequency of a component in a harmonic series is less than the size of the sieve, it remains part of the series and contributes to the virtual pitch. Frequencies lying outside of the sieve's boundary are heard as separate sounds (pitches) apart from the virtual pitch. The data cited above suggest that the sieve width is somewhere between 3% to 8% of the frequency of any one

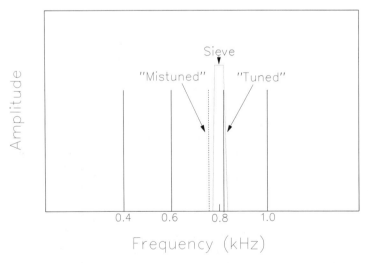

FIGURE 6.5. A schematic diagram indicating the concept of a spectral sieve for detecting the presence of a mistuned harmonic in a harmonic series. If a spectral component falls within the boundaries of a sieve opening, it will be incorporated into the harmonic series and add to the perception of the complex pitch. If a spectral component falls outside of the sieve boundary, it will segregate from the rest of the complex and be heard as a pitch that is different from that of the complex.

component in the spectrum. A combination of the sieve theory and one of the other theories of pitch perception (e.g., Goldstein 1973; Terhardt, Stoll, and Seewann 1982) provides a good account for how spectral components can be combined into a single auditory percept, virtual pitch. Perceiving the pitches in a complex sound field is one means of determining sound sources.

Harmonicity has also been shown to affect spectral grouping for speech segregation. Voiced speech is a harmonic stimulus spectrally shaped by the resonances (formants) of the vocal tract. A common fundamental frequency can enhance the fusion of formants for speech perception (Broadbent and Ladefoged 1957; Darwin 1981; Gardner, Gaskill, and Darwin 1989). Darwin and Gardner (1986) found that mistuning of 3% to 8% will exclude a single harmonic from a vowel sound as measured by phoneme classification. This result is consistent with other estimates of the variance of a harmonic sieve for complex pitch perception. A difference in fundamental frequency aids the identification of two concurrent vowels (Scheffers 1982; Zwicker 1984; Summerfield and Assmann 1991) and also enhances the segregation of competing speech messages (Brokx and Nooteboom 1982). In the two-concurrent vowel identification task, both fundamental frequency separation and differences in spectral profile between the two vowels affect performance (Scheffers 1982; Zwicker 1984). Even when both vowels share a common fundamental frequency, differences in their spectral profile can allow for the identification of two speech sources.

4.1 Models of Pitch Perception

Historically, there have been two general forms for the theories of pitch perception: spectral pitch models and temporal pitch models (see Moore, Chapter 3, for the description of a particular model). Both classes of models assume that the components of a complex sound are resolved by the frequency selectivity of the auditory periphery. The models differ in the form of pattern recognition that is used to extract a pitch prediction from the spectrotemporal information at the output of the auditory periphery. In general, the spectral models attempt to find the best fitting harmonic series to the spectral components of the sound as they are represented in the auditory periphery. The fundamental of the best fitting harmonic series indicates the pitch of the complex sound. The temporal models assume that the auditory system searches for robust periodicities in the temporal neural pattern at the auditory periphery, and the reciprocal of that period is the pitch of the sound. Figures 6.6 and 6.7 depict the general ideas of these two modeling approaches for a harmonic series of tones yielding a pitch of 200 Hz and an inharmonic series which typically yields two somewhat ambiguous pitches: approximately 180 and 220 Hz. The "best fitting" harmonic series and the most robust periodicity in the waveforms are shown for each stimulus. The various spectral models differ in the details of how the best fitting harmonic series is arrived at (Fig. 6.6). A number of different methods for extracting the period-

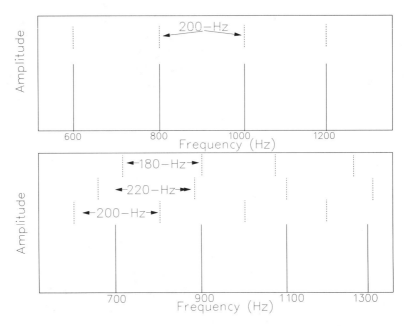

FIGURE 6.6. A schematic diagram summarizing the basic concept of many spectral models of pitch. The top panel shows a harmonic series with a fundamental of 200 Hz. The fundamental of the best fitting harmonic series corresponds to a pitch of 200 Hz. In the bottom panel, the series of tones are placed 200 Hz apart but are not harmonics of 200 Hz (the components are at 900, 1100, 1300, and 1500 Hz). Two harmonic series come closest to matching the spectrum of the sound, one with a fundamental of 180 Hz and the other with a fundamental of 220 Hz, indicating the two pitches often perceived for this complex sound.

icity of a waveform have been used to form the different temporal models of pitch (Fig.6.7). Although the two modeling approaches are different in their assumptions regarding pitch extraction, they are often equivalent in that a particular temporal approach is a simple transform of a particular spectral approach. For instance, autocorrelation is one means of determining the periodicities in a waveform. However, autocorrelation (a type of temporal process) and power spectrum (as would exist in a spectral model) are Fourier transforms of one another (see Wightman 1973a,b). Thus, it is often difficult to determine which approach (temporal or spectral) is the better method for modeling pitch perception. Since many of the models lead to exact but differing predictions for particular stimulus conditions, these models have and can be tested (see de Boer 1976). At present, no one model has emerged as the best one to account for the wide variety of stimulus conditions that yield virtual and spectral pitches (see de Boer 1976; Plomp 1976).

The current versions of some pitch models recognize that the pitch of a complex sound is primarily determined by the lower harmonics (the first four

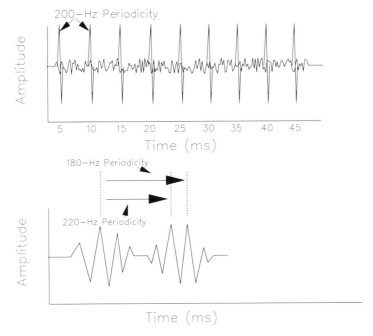

FIGURE 6.7. A schematic diagram summarizing the basic concept of many of the temporal models of pitch. In the top panel, the temporal waveform has a 200-Hz periodicity (it was generated with the harmonics of 200 Hz all presented in cosine phase). The 200-Hz periodicity accounts for the 200-Hz pitch. The bottom panel is a schematic diagram of a complex made up of tones spaced 200 Hz apart but not at harmonics of 200 Hz (i.e., similar to that described in the bottom panel of Figure 6.6). When this sound is analyzed in certain spectral regions, there are two prominent periodicities between the major peaks in the time-domain waveform, one with a frequency of 180 Hz and one with a frequency of 220 Hz (see Plomp 1976).

to six) of the fundamental that can be spectrally resolved, thus emphasizing a spectral approach to modeling pitch (Moore 1989). However, most of these versions of pitch models assume some form of temporal analysis is used to extract a pitch estimate from the resolved harmonics. Thus, these newer modeling efforts represent a combination of the earlier temporal and spectral approaches to modeling complex pitch.

5. Spatial Separation

In 1953 Cherry coined the term "cocktail party effect" to describe the ability to determine the sources of sounds in a multisource acoustic environment. Cherry was primarily interested in what he called the "one-world, two-ear problem." That is, we have but one acoustic stimulus (the complex sound

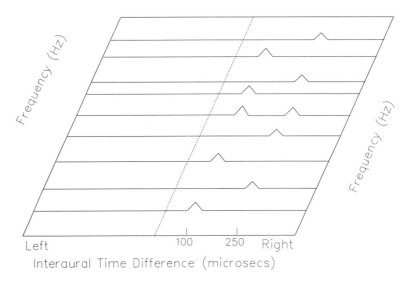

FIGURE 6.8. A two-dimensional schematic diagram indicating how two sound sources, one made up of a number of spectral components presented with a 100-μs interaural time difference (ITD) and the other source consisting of a number of components presented with a 250-μs ITD, might be represented. Some additional processes are required in order for the auditory system to fuse components based on a common ITD. Such a process might have to deal with a situation in which one or more spectral components are common to both sources, as indicated in this figure.

field in Yost, Chapter 1, Fig. 1) and we use our two ears to sort out the sound sources that constitute this acoustic stimulus. Cherry, therefore, argued that it is our ability to spatially locate sounds that plays a major role in the cocktail party effect. Thus, the interaural differences of time and level might provide powerful cues for sound source determination. Figure 6.8 is a schematic diagram indicating the problem facing the binaural system in using interaural differences to determine the sources of sound. How does the system recognize that the spectral components having 100 μs of interaural time difference are one source at one location and those having 250 μs of interaural time difference belong to another source at a different location?

Cherry (1953) used what has been called a selective attention paradigm to investigate sound source determination for stimuli presented to two ears. In most of his experiments, a sound or message was presented to one ear (test ear) while a competing sound was presented to the other ear, either at the same time or with some differences in the timing of the information arriving at the two ears. The listener's task was to determine some aspect of the sound or message presented to the test ear (e.g., to report the message). In general, separating the messages between the ears allows the listener to process the test message better than when both sounds are mixed and presented to only one ear. However, these selective attention experiments performed with headphones do not reflect the actual acoustics that occur in a real-world listening situation where both ears would receive all of the sounds and interaural differences would be used to separate one source from another. There have not been many studies of selective attention or the "cocktail party effect" in real listening environments.

5.1 The Masking Level Difference

In the simplest "cocktail party," there is a signal at one location and a background noise at a different location. To a large extent, this stimulus paradigm describes a number of experiments (see Green and Yost 1975; McFadden 1975 for reviews) involving the masking level difference or the MLD (sometimes called the binaural masking level difference or the BMLD). In a MLD experiment, a signal with one set of interaural differences is added to a masker with a different set of interaural differences. Signal detection thresholds are then measured as a function of the interaural parameters. (A few experiments measured thresholds for discriminating some change in the signal, e.g., frequency discrimination, or the ability to recognize one signal from a set of possible signals, e.g., speech recognition.) Wightman and Kistler (Chapter 5) showed that changes in the interaural phase and intensity of a sound result in the sound having a different lateral location when presented over headphones. When presented individually, a signal and masker that differ in interaural parameters would be lateralized at different positions. However, it should be pointed out that the stimulus paradigm calls for the signal plus masker complex to be contrasted with the masker alone presentation. The effect on interaural differences of the addition of the signal to the masker compared to those existing for the masker alone is illustrated in the vector diagrams of Figure 6.9.

Table 6.1 depicts some of the stimulus conditions that have been used in the study of the MLD and the typical signal-to-noise ratios determined at threshold when the signal is a 500-ms, 500-Hz tone masked by a continuous, broadband noise of moderate level (i.e., approximately 30 to 40 dB spectrum level). The signal-to-noise ratio required for threshold performance is lower when the stimulus condition is dichotic than when it is diotic or monotic while the diotic and monotic thresholds are the same. The difference in the thresholds between a dichotic and the monotic (or diotic) condition is the MLD in decibels as shown in Table 6.1. The size of the MLD is dependent on the frequency of the signal. The MLD is largest for signal frequencies of 300 to 700 Hz. The MLD has been investigated for a large number of other stimulus conditions (see Green and Yost 1975; McFadden 1975 for reviews), and this literature clearly shows that, for many stimulus conditions, both the detectability, and to a lesser extent the discriminability and recognition of signals, improves when the signal is presented with a different set of interaural values than the masker. This improvement in performance due to spa-

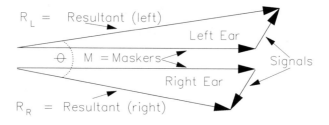

Interaural Phase Difference = θ
Interaural Level Difference = $R_L - R_R$
Monaural Level Difference = R_L (or R_R) − M

FIGURE 6.9. A vector diagram indicating that, in a Masking-level difference (MLD) paradigm, the signal and masker are combined into one stimulus that has an interaural time difference and an interaural level difference. The difference between the masker-plus-signal and the masker alone presents a monaural level difference. All three differences are potential cues for detection. However, when the signal level is low relative to that of the masker, the interaural differences determine signal detection.

TABLE 6.1. The MLD in dB for a variety of stimulus conditions.

Interaural Condition Compared to MmSm	MLD
MmSm, MoSo, MπSπ	0 dB
MπSm	6 dB
MoSm	9 dB
MπSo	13 dB
MoSπ	15 dB

So: signal presented binaurally with no interaural differences (diotic)
Mo: masker presented binaurally with no interaural differences (diotic)
Sm: signal presented to only one ear
Mm: masker presented to only one ear
Sπ: signal presented to one ear 180° out-of-phase with the signal presented to the other ear
Mπ: masker presented to one ear 180° out-of-phase with the signal presented to the other ear

tial separation is consistent with the "cocktail party effect" as one aspect of sound source determination.

Most of the MLD literature involves stimulus presentation through headphones. A few studies have investigated the detection and recognition of signals coming from one loudspeaker while masking or interfering stimuli were presented from loudspeakers at different locations (Plomp and Mimpen 1981; Saberi et al. 1991). In these free-field cases, signal detection threshold can be as much as 10 dB lower when the masker and signal are spatially separated than when they emanate from the same sound source. The greater the spatial separation between the masker and the signal, the easier it is to hear the signal.

5.2 Other Binaural Phenomena and Auditory Perception

It is clear from the vector diagrams of Figure 6.9 that when the signal and masker have different interaural values their addition produces an interaural phase and intensity difference. What happens if there is *only* an interaural phase difference or an interaural intensity difference in some region of the spectrum of a complex sound such as a noise? This condition could produce the amplitude and phase spectra in Figure 6.10 if the phase and intensity differences were introduced in only one spectral region. Note that, unlike in the MLD procedure, there is only an interaural phase shift or an interaural amplitude difference; no other interaural or monaural differences are present. If the spectral region over which the phase is shifted or the amplitude altered is relatively narrow, then listeners report hearing a weak pitch associated with the region of the spectrum that is interaurally altered. The perceived pitch is lateralized away from the midline where the background noise is lateralized. These dichotically produced pitches are referred to as Huggins pitches after Cramer and Huggins (1958) who first demonstrated the existences of these pitches (see Moore, Chapter 3, for a brief discussion of Huggins pitch as it relates to models and theories of pitch perception). The ability to hear the Huggins pitch must be due to binaural processing since there are no monaural cues that can be used (Yost, Harder, and Dye 1987). The existence of the Huggins pitches demonstrate a form of analytic listening in which the auditory system uses the common interaural differences of the altered segment of noise to separate those spectral components from the rest of the waveform that has a different set of interaural parameters (in the example shown above, no interaural differences).

So far we have primarily described experiments involving the detection of signals. A number of studies have investigated the discrimination and recognition (especially speech recognition) of signals under dichotic listening conditions, as compared to diotic or monotic conditions (see Green and Yost 1975 for a review). In general, these studies showed that discrimination and recognition are better when the signal has a different set of interaural values than a background or masking sound (the conditions are dichotic), but only

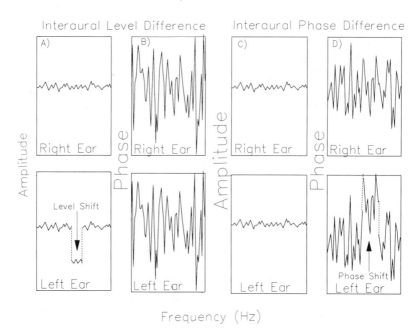

FIGURE 6.10. The spectra (amplitude—panels A and C; and phase—panels B and D) of complex, dichotic sounds presented with only an interaural phase difference (right two panels—panels C and D) or an interaural level difference (left two panels—panels A and B). The amplitude spectrum (panels A and C) and phase spectrum (panels B and D) are altered at the left ear (bottom row—left ear; top row—right ear) in a limited region of the spectrum. Under the appropriate conditions listeners report hearing a pitch (referred to as the Huggins pitch) associated with the band of noise that is interaurally altered.

when the signal is difficult to detect. As the signal level increases and it becomes easier to detect, the differences between dichotic and diotic discrimination and recognition performance become small or disappear.

A few studies have also investigated the localization and lateralization of a target signal in the presence of other spatially separated stimuli. Listeners do not appear to be as sensitive at processing interaural time and level differences of a target stimulus when it is presented with other stimuli (Dye 1990; Trahiotis and Bernstein 1990; Buell and Hafter 1991; Woods and Colburn 1992). Figure 6.11 shows the result of a study by Stellmack and Dye (1993) in which the task was to discriminate a change in the interaural time difference of a 753-Hz target tone in the presence of other tones (distractor tones) with no interaural differences. The number of distractor tones and the frequency spacing between tones were the two independent variables. The dashed line at the bottom of the figure indicates the interaural time difference threshold for the 753-Hz target tone presented by itself. Interaural time difference

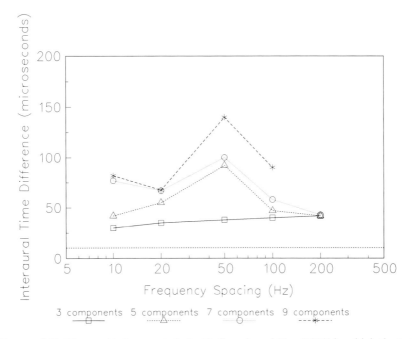

FIGURE 6.11. The results from a study by Stellmack and Dye (1993) in which the task was to detect an interaural time difference of a 753-Hz target tone in the presence of diotic distractor tones. The number of tones in the complex and their frequency spacing were varied. The horizontal line across the bottom of the figure indicates thresholds for detecting an interaural time difference when the target was presented alone. The interaural time difference thresholds were always higher when the diotic tones were present. (From Stellmack and Dye 1993 with permission).

thresholds for the target rose when the diotic tones were added, with the threshold increment increasing with the number of distractor tones. Substantial interference was found for wide spectral separations between the tones. These results suggest that, in some conditions, the auditory system may not be analytic in its ability to use disparities in interaural values across the spectrum of a complex sound to separate one sound source from other sources.

6. Temporal Separation

The first four variables discussed in Sections 2–5 share the common characteristic of representing static or constant differences among the outputs of multiple sound sources. The remaining two variables, temporal separation and modulation, rely on distinct time-varying characteristics of the individual sources. The outputs of most sound sources are temporally complex, consisting of a series of acoustic events, with each event often varying in some manner over its duration. When there are multiple sound sources, their outputs are almost always asynchronous, beginning and ending at different points in time. The physical variable temporal separation refers to this asynchrony, involving situations in which the acoustic events originating from individual sources are either concurrent or nonoverlapping.

With asynchronous sources, there are two aspects to sound source determination. The first is the detection and processing of the individual acoustic events that arise from each of the multiple sources. There has been extensive psychoacoustical study of the effects of temporal separation and asynchrony between a probe signal and masker on signal detection (see Viemeister and Plack, Chapter 4). The second aspect concerns the perceptual organization of the sequence of acoustic events generated by a multisource sound field. The listener must determine if separate events originated from the same or different sources. This sequential organization, termed stream segregation by Bregman and his coworkers (e.g., Bregman and Campbell 1971; Dannenbring and Bregman 1978), involves the processing of a series of acoustic events, attributing them to one or more implied sources.

The duration of the temporal separation between the onsets of a probe signal and masker affects signal detection thresholds in both simultaneous and nonsimultaneous masking conditions. In general, the effect of temporal separation is most evident for short-duration probe signals. [However, McFadden and Wright (1992) reported that, in a simultaneous masking experiment with multiple masker bands, a similar effect was found for temporal separation for 50- and 240-ms probe signals. For both probe durations, individual subjects showed comparable improvement in probe detection thresholds with increasing temporal separation.] That the duration of most sound source outputs extends beyond the typical range of the psychophysically measured effects of temporal separation might suggest a limited relevance to source determination. However, the psychophysical work is important in suggesting aspects of neural processing that affect source determination. Results from masking studies suggest involvement of neural adaptation and a role of signal onset in determining threshold. Both aspects may directly relate to sound source determination.

6.1 Adaptation and Auditory Enhancement

Many of the temporal effects observed in psychophysical masking procedures are modeled through the involvement of peripheral neural adaptation (see Viemeister and Plack, Chapter 4). In the forward masking procedure, where the probe and masker do not temporally overlap, the masker adapts fibers tuned to the probe frequency. Increasing either the frequency or temporal separation between the probe and masker reduces the amount of masking. When the masker still precedes the probe but the two completely overlap in time, the dependency of temporal effects on spectral characteristics shows a more complex pattern. Masking overshoot (see Viemeister and Plack,

Chapter 4) is the increase in probe detection threshold, with the probe onset near the masker onset. That the overshoot increases with increasing masker bandwidth beyond the signal critical bandwidth and is also reduced by preceding the masker with an off-signal frequency noise band indicates both on- and off-signal frequency effects (Zwicker 1965; McFadden 1989; Bacon and Smith 1991; Carlyon and White 1992). While the on-frequency component in masking overshoot may, at least in part, reflect peripheral adaptation (Smith and Zwislocki 1975), the off-frequency component indicates involvement of either more central adaptation (McFadden 1989) or the synchronized neural response across many nerve fibers to the masker onset (Champlin and McFadden 1989; Bacon and Smith 1991).

One consequence of adaptation is to enhance the neural response to spectral variation in the stimulus. With asynchronous sound sources, adaptation of the response to the initial source could enhance the relative response to subsequent sources. Enhancement due to prior stimulation can be evident as a "negative afterimage" (Zwicker 1964; Wilson 1970). After listening to a broadband, adapting stimulus with spectral notches or ripples and then to the stimulus without the notches (test stimulus), listeners often report the perception of a pitch or timbre in the test stimulus that corresponds to the spectral regions of the notches in the adapting stimulus (see Fig. 6.12). One

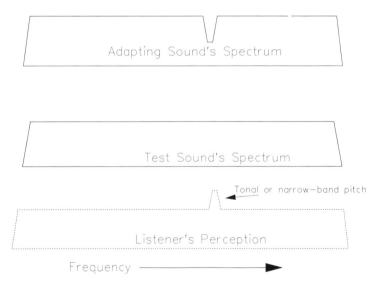

FIGURE 6.12. A schematic diagram of one procedure used to study auditory enhancement. An adapting stimulus which contains a spectral notch is presented first. This presentation is immediately followed by a test stimulus which does not contain the notch (i.e., the test stimulus has a flat spectrum). Under the appropriate conditions, the listener detects a pitch during the test presentation which corresponds to the spectral region of the notch in the adapting stimulus.

explanation for the perception of the pitches and timbres assumes differential adaptation to the initial stimulus, allowing for the enhanced representation of spectral components in the test stimulus. Auditory enhancement due to prior stimulation has been studied with masking, discrimination, and identification procedures (Viemeister 1980; Viemeister and Bacon 1982; Summerfield, Sidwell, and Nelson 1987; Carlyon 1989; Summerfield and Assmann 1989. As measured by either masking effectiveness or loudness matching, the enhancement is not simply based on the reduced sensitivity of adapted regions but appears to represent a gain in the response from unadapted regions (Viemeister and Bacon 1982; Summerfield, Sidwell, and Nelson 1987). This result led Viemeister and Bacon (1982) to speculate that the initial stimulus adapts the suppression that would normally exist for the test stimulus.

While auditory enhancement effects appear related to temporal effects in masking, certain aspects of some enhancement phenomenon cannot be accounted for solely by the involvement of peripheral neural adaptation. The growth and decay of the enhancement effect can extend past the time span generally associated with peripheral adaptation (Wilson 1970; Viemeister 1980). Also, a 30-dB range of level variation of the initial stimulus does not have a significant effect, as measured by a change in masking of the subsequent stimulus (Carlyon 1989). Overall, the effects can be thought of as emphasizing changes in a spectral profile due to newly arriving information. Perhaps the simplest example of this situation is the detection of one source that produces an onset asynchrony among multiple sound sources.

6.2 Onset and Offset Asynchronies

Onset and offset asynchronies can aid the perceptual decomposition of a complex stimulus into harmonic subsets, influence the contribution of a harmonic to the phonetic quality of a vowel, and affect the tendency of a single spectral component of a series of acoustic events to either become part of a sequential stream or fuse with other simultaneously present components. Rasch (1978) measured the masked thresholds for identifying the direction of pitch change of a complex tone. The masker was also a complex tone whose pitch was below that of the target. If the masker and target were synchronous, the target-tone intensity was in the range of -20 to 0 dB (re: masker level) at threshold, with performance improving as the masker and target were mistuned. Thresholds dropped to roughly -60 dB with a 30-ms onset asynchrony between target and masker. In conditions with an onset asynchrony, masker level, reverberation, or temporal overlap between the target and masker had little effect on target-tone thresholds. With the target onset preceding the masker onset, the procedure measured the backward masking of complex pitch discrimination.

Darwin and his coworkers have measured the effects of temporal asynchrony on the contribution of an asynchronous component to the perception of a complex sound. These procedures do not measure thresholds for temporal masking but more directly evaluate spectral fusion. Darwin and Ciocca (1992) investigated the influence of an onset asynchrony on the contribution of a mistuned harmonic to the pitch of a complex tone. With either ipsilateral or contralateral presentation, the contribution of the mistuned component to the pitch of the complex progressively diminished with increasing onset asynchrony, from roughly 80 to 320 ms. In work with speech stimuli, the effect of asynchrony was apparent with shorter onset disparities. A 32-ms onset or offset asynchrony of a single harmonic of a vowel sound was found to significantly alter the phoneme boundary, with the change in vowel quality continuing to change as the asynchrony was increased to 240 ms (Darwin 1984; Darwin and Sutherland 1984). Though the effect of a 32-ms onset asynchrony was greater than that of a 32-ms offset disparity, the effect of an offset disparity indicates involvement of a process other than adaptation. The reduced contribution to vowel quality of a harmonic that begins before the complex was countered by adding a second asynchronous tone to the harmonic. Darwin and Sutherland (1984) speculated that the two-tone precursor formed a separate perceptual group that ended at the vowel onset, allowing the one harmonic to then contribute to the vowel. This kind of interaction between competing structuring of a temporally complex stimulus has received greatest attention in studies of auditory stream segregation.

6.3 Stream Segregation

Bregman and his coworkers (Bregman and Pinker 1978; Dannenbring and Bregman 1978) have investigated the role of onset asynchrony in stream segregation. An auditory stream refers to the perception of a series of sounds as a single unitary sequence, in essence, the output of a single sound source. Most experiments on stream segregation evaluate whether a repeating series of sounds forms a single or multiple stream sequences (see Fig. 6.13). When using tonal stimuli, the perception can be one of hearing one or several concurrent melodies. In the simplest case, the stimulus is a repeating series of two pure tones with the frequency separation of the tones and the repetition rate the primary experimental variables. Increasing the value of either of these variables increases the segregation of the two tones into separate streams (Miller and Heise 1950; Bregman and Campbell 1971; van Noorden 1975; Dannenbring and Bregman 1976). The absolute values of the frequency separation and repetition rate at the threshold for stream segregation can vary with the metric used to determine threshold and the requirements of the task. In fact, if listeners are asked to judge the frequency separation needed to just be able to focus attention on a single stream, the threshold values are roughly independent of repetition rate (van Noorden 1975).

Along with frequency separation and repetition rate, stream segregation is affected by the other physical variables considered in this chapter on source determination. A sequence of a single pure tone alternating in level by as little as 3 dB can segregate into two streams (van Noorden 1977). With complex

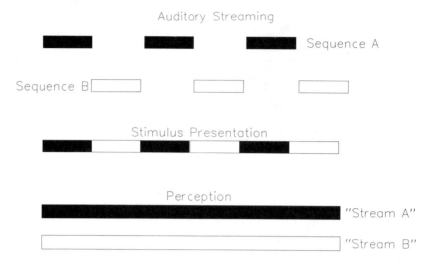

FIGURE 6.13. A schematic diagram indicating the type of procedure used in many streaming experiments. Two sounds (e.g., two pure tones differing in frequency sequence A vs sequence B) are presented as alternating sequences. Depending on stimulus conditions and task parameters, listeners report hearing either a single stream that alternates in pitch or two streams each with its own pitch (as is shown as Perception in the figure).

stimuli, differences in the spectral profile can lead to stream segregation. As an extreme example, a sequence of alternating noise and tone bursts segregates (Dannenbring and Bregman 1976). When using complex tonal stimuli with a common fundamental, stream segregation can be achieved by varying the harmonic number or amplitudes (Singh 1987; Bregman 1990; Bregman, Liao, and Levitan 1990). Conversely, fixing the spectral profile while varying virtual pitch can also segregate tonal sequences. Coherence of either amplitude or frequency modulation affects streaming (Chalikia and Bregman 1989; Bregman, Levitan, and Liao 1990). Finally, spatial separation can affect the grouping of a sequence of sounds (Judd 1979; Steiger and Bregman 1982; Kubovy 1987; Barsz 1991).

Another set of studies have investigated the perception of sounds that occur in succession. The paradigm for these studies is similar to that used in many studies of streaming: two sounds (a masker and a signal) are presented in an alternating sequence. When the signal and masker excite overlapping regions of the auditory periphery and the signal level is low compared to that of the masker, the signal is often perceived as a continuous sound (e.g., Houtgast 1972; Warren, Obusek, and Ackroff 1972; Verschuure 1981) . This illusion of continuity has also been obtained with complex signals such as tone glides and speech (e.g., Bashford and Warren 1987; Ciocca and Bregman 1987). A consequence of the illusion can be to maintain source continuity among competing sources.

Results from streaming studies return to the issue of the competition between the sequential and simultaneous organization of a temporally complex auditory stimulus. This interaction can be observed with diotic stimuli in which a repeating sequence contains three events, a single event followed by two temporally overlapping events (Bregman and Pinker 1978; Dannenbring and Bregman 1978; Steiger and Bregman 1982). Onset and offset synchronies, along with harmonicity, promote integration or fusion of the temporally overlapping events. Asynchrony between the two concurrent events or a small frequency between the single event and one of the two following events weakens the fusion to encourage sequential streaming. These results are consistent with processing that attempts to factor a complex auditory input according to potential or implied sources.

6.4 Informational Masking

An early observation in streaming experiments was the loss of information regarding sequential events in segregated streams (Bregman and Campbell 1971). The threshold for detection of a change in an single event of a sequence is highly dependent on the amount of information contained within the

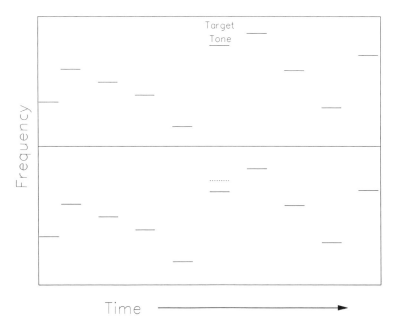

FIGURE 6.14. A ten-tone pattern (each short line represents one tone with a particular frequency presented at a particular time in the pattern) typically used to study informational masking. The listener's task is to determine if the frequency of the target component changes in a same-different task (i.e., either the same ten-tone pattern is presented twice or a slight change in target frequency is introduced in the second presentation of the pattern, as shown in the bottom pattern).

sequence (Watson and Kelly 1981). Variation in the target location within a pattern, the total number of elements in the pattern, and in stimulus uncertainty affect the information content of the sequence. The inverse relationship between target resolution and pattern complexity has been termed informational masking (Pollack 1975; Watson and Kelly 1981).

Watson and his colleagues (Watson et al. 1975; Watson 1976; Watson and Kelly 1981) have conducted a large series of studies evaluating detection and discrimination for a single event in a series of successive sounds (see Fig. 6.14). In a typical experiment, the sounds are a sequence of ten tones differing in frequency. In a same-different psychophysical procedure, the task is to determine if the frequency of one tone has been changed. The frequency range over which the ten tones may vary, the duration of each tone, and the level parameters are often chosen to reflect those that occur in speech. The sounds are not perceived at all like speech, but Watson and colleagues argued that, by using sounds whose temporal and spectral properties are like those associated with speech, they might be able to determine the auditory acuity for the physical variables that constitute a speech sound (Watson et al. 1975).

A number of important principles have been uncovered by this work on complex pattern perception. In the earlier research, the listeners were asked to determine if a frequency (or level or duration) change in the target sound had occurred when, from trial to trial, the frequency of each tone in the entire ten-tone sequence was randomly changed. Figure 6.15 shows the results from such experiments. Frequency discrimination performance was very poor, with thresholds highest for low-frequency tones occurring at the beginning of the pattern.

In follow-up experiments (Watson and Kelly 1981), the uncertainty associated with randomly varying the frequency of the ten tones was reduced in a variety of ways (e.g., reducing the range over which the frequencies of the tonal components might vary). Watson and Kelly (1981) termed this a reduction in "stimulus uncertainty." In a minimal uncertainty experiment, the listener would receive the same ten-tone pattern on every trial, with only the frequency of the target tone changing in the same-different paradigm. As the level of uncertainty decreased, discrimination ability improved, most notably for target tones located toward the beginning of the pattern. Watson and Kelly (1981) described the increase in discrimination thresholds that occurred due to an increase in stimulus uncertainty "informational masking" to differentiate this from a loss of sensitivity that depends on auditory processing, such as that which occurs at the auditory periphery. Thus, the loss of sensitivity for discriminating spectral changes at the beginning of a complex pattern were largely attributed to informational masking.

Another finding from the research on complex tonal-pattern perception involves the ability of listeners to process one segment of the pattern as a function of the proportion of the total pattern that this segment occupies. Kidd and Watson (1992a) have demonstrated that the sensitivity for pro-

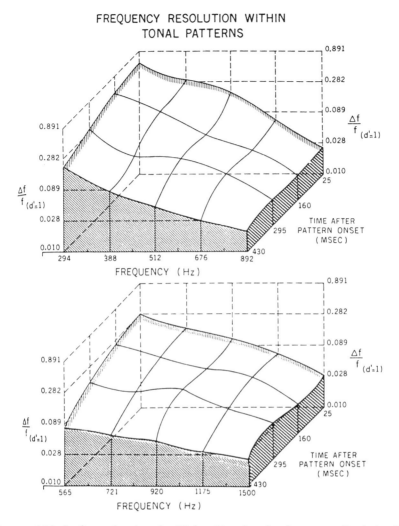

FIGURE 6.15. Surfaces showing the Weber Fraction for frequency discrimination ($\Delta f/f$) for a d' of 1.0 for the target tones in the ten-tone patterns shown in Figure 6.14. The two surfaces represent two different ranges of frequencies (top figure: 294 to 892 Hz; bottom figure: 565 to 1500 Hz). Discrimination is poorest for low-frequency tones presented near the beginning of the pattern. Performance in this high-uncertainty procedure is very poor as compared to the Weber Fraction of 0.003 often obtained for frequency discrimination of pure tones presented in isolation. (From Watson et al. (1975) with permission.)

cessing a frequency change of one tone (the target) is determined by the proportion of the total time of the complex pattern occupied by the target tone. This proportionality appears to hold despite a number of changes that may be introduced to the entire complex pattern. These and other findings (Kidd and Watson 1992b) have led Kidd and Watson to formulate the proportion-of-the-total-duration (PTD) rule for determining how the sensitivity for processing one temporal segment of a complex sound depends on the total duration of the sound.

7. Coherent Temporal Modulation

McAdams (1984) and Bregman (1990) have argued that coherence of change among spectral components is used by the auditory system to identify those components as originating from a single source. This change or modulation may refer to variation in either the frequency or amplitude of the components. With amplitude modulation (AM), the instantaneous amplitude of the carrier varies in proportion to the modulating signal. With frequency modulation (FM), the instantaneous frequency of the carrier varies in proportion to the modulating signal. Modulation is characteristic of many noise signals. If the noise is bandlimited so its spectrum covers a narrow frequency range, the noise can be represented as a carrier whose amplitude and frequency are randomly modulated. Due to the nature of many sound-producing mechanisms, the pattern of modulation represents a fundamental and distinguishing characteristic of the sound source output.

The temporal and spectral characteristics of sinusoidal AM are discussed by Viemeister and Plack (Chapter 4). In the case of sinusoidal FM, the modulated signal, $s(t)$, has the form $s(t) = \cos[2\pi f_c t + \beta \cos(2\pi f_m t)]$, where t is time, f_c is the carrier frequency, f_m is the modulation frequency, and β is the modulation index. With Δf the peak deviation of the instantaneous frequency of the FM signal from the carrier frequency, $\beta = \Delta f/f_m$. Unlike sinusoidal AM where there are only two side bands in the spectrum of a modulated pure-tone carrier, the spectrum of an FM wave contains an infinite set of side bands located symmetrically about the carrier frequency, with frequency separations at integer multiples of the modulation frequency [e.g., $(f_c \pm f_m)$, $(f_c \pm 2f_m)$; see Fig. 6.16]. If β is small compared to unity, only the side band pair closest to the carrier has significant amplitude. In general, over 98% of the power with sinusoidal FM is contained within a bandwidth of $2(\Delta f + f_m)$.

If coupled to resonators (e.g., the formant characteristic of the human vocal tract) or processed through filters (e.g., the filtering of the peripheral auditory system), the frequency changes of an FM signal can result in AM of the system output. When driving multiple filters differing in center frequency, the amplitude modulation of the filter outputs is cross-spectrally incoherent. An extreme example of this type of FM to AM conversion is shown in

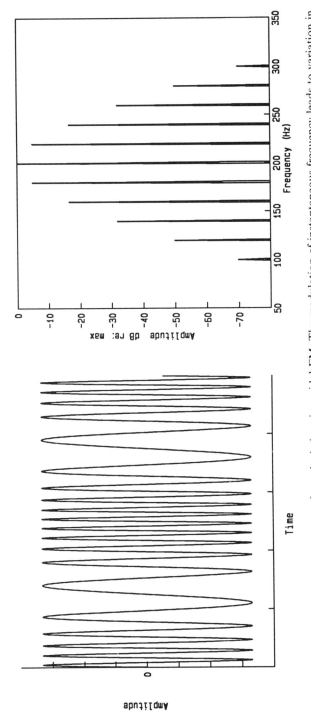

FIGURE 6.16. On the right is a time waveform depicting sinusoidal FM. The modulation of instantaneous frequency leads to variation in the time between successive zero crossings over the signal duration. The left panel shows the amplitude spectrum of a 200-Hz pure-tone carrier frequency modulated at 20 Hz with β equal to 1.0.

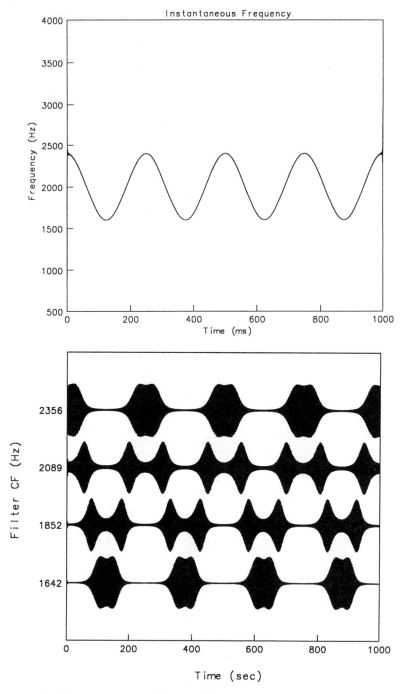

FIGURE 6.17. The response of the filter bank stage of the Patterson and Holdsworth (Patterson et al. 1992) model to a frequency-modulated pure tone. The input signal is a 2-kHz carrier modulated at 4 Hz with β equal to 100. The modulation of the instantaneous frequency of the carrier is shown in the top panel. The bottom panel shows the response of four simulated filters. Due to the sweeping of the signal through the filter passbands, filter outputs are amplitude modulated.

Figure 6.17. The amplitude modulation of the four filter outputs is incoherent not only in terms of envelope phase, but also in the pattern of amplitude modulation.

7.1 Frequency Modulation

The example given in Section 1 from McAdams' (1984) work illustrates the effect of coherent FM, as exemplified by vibrato or jitter, to allow for sound source segregation. Other studies have shown that, for stimuli in which a subset of spectral components is coherently modulated to represent a source, judgments of source prominence (McAdams 1989; Marin and McAdams 1991), the number of apparent sources (Chalikia and Bregman 1989; Gardner, Gaskill, and Darwin 1989), and stream segregation (Bregman and Doehring 1984) are higher than without modulation. However, when several subsets of components are modulated, coherent FM across subsets does not diminish the prominence of a target subset (McAdams 1989; Marin and McAdams 1991). That is, coherence may help segregate one subset from a complex, but it does not always lead to fusion of different subsets. McAdams (1989) suggested that the presence of harmonic relationships only within, but not across, subgroups allows for the retention of target prominence, despite coherence of modulation with other nontarget components.

Results from phoneme identification procedures do not indicate involvement of coherent FM in source segregation. Demany and Semal (1990) measured masked thresholds for the identification of vowels. Addition of vibrato as 6-Hz FM of the vowel fundamental lowered thresholds only for naive subjects and when the masker was a tone. Also, incoherence due to phase or rate manipulation of the FM of a single harmonic of vowel does not change the phonetic quality of the vowel; the incoherently modulated harmonic is still perceived as if it were integrated into the vowel spectrum (Gardner and Darwin 1986; Gardner, Gaskill, and Darwin 1989). Comparing the effect of static mistuning (the steady-state mistuning of a frequency component in the complex) to FM modulation incoherence, Gardner, Gaskill, and Darwin (1989) argued that the effect of FM incoherence is mediated by the mistuning it introduces as the instantaneous frequency of components varies with differing trajectories. Carlyon (1991) made a similar argument based on the thresholds for discriminating between coherent and incoherent FM of complex tones. In his work, subjects were unable to detect FM incoherence with inharmonic stimuli. With harmonic stimuli for which mistuning due to FM incoherence becomes a reliable detection cue, incoherence was detectable. Carlyon argued that the involvement of FM coherence in spectral segregation does not rely on a process specific to the detection of FM coherence but is instead mediated by covarying cues of harmonicity or envelope coherence. As discussed in Section 7, FM can lead to amplitude modulation of the output of the filter bank of the auditory periphery. Wakefield and Edwards (1989) modeled the detection of FM coherence by the cross-spectral AM disparities generated as the frequency-modulated carriers sweep through peripheral auditory filters.

7.2 Amplitude Modulation

Involvement of envelope coherence in spectral segregation was first suggested by Helmholtz (1885/1954). Stimulus onset and offset discussed in Section 6.2 represent one aspect of the temporal envelope of a sound. Variation in the temporal envelope over the duration of a sound is amplitude modulation. If the amplitude of a single component of a tonal complex is either briefly dropped or continuously modulated, that component tends to stand out from the complex (Kubovy and Daniel 1983; Kubovy 1987). Such amplitude changes are a form of AM. Pitch segregation by a momentary amplitude disparity depends on the frequency spacing between the target and nontarget components, with the effect diminishing if adjacent components are separated by less than a critical bandwidth. Sheft and Yost (1989a) and Hall and Grose (1991) have shown that the discrimination of tonal complexes based on which component is modulated requires an AM depth well above the AM detection threshold. This result is perhaps troublesome for speculation on the involvement of envelope modulation in sound source segregation. Proper spectral segregation based on AM requires associating each distinctive modulation pattern with its appropriate carrier. When source outputs contain multiple components, listeners must detect the envelope coherence among the components of the individual sources and then, in some manner, selectively process the various fluctuation patterns that characterize different sources.

Results from two studies by Sheft and Yost (1990, 1992a) demonstrated the requisite selectivity in processing of AM. The first measured the ability to detect envelope coherence among target noise bands in the presence of masking bands. The envelope of narrowband noise fluctuates with an irregular pattern. For ideal narrowband noise, the expected number of maxima of the envelope is roughly 0.64 times the spectral bandwidth (Rice 1954). The power spectrum of the envelope is a continuous sloping function, ranging from the dc to frequencies beyond the noise bandwidth (Price 1955). Experimental conditions used noise bandwidths ranging from 12.5 to 200 Hz. A cued two-interval, forced-choice (2IFC) procedure was used; each trial was preceded by presentation of the synchronous target noise bands as a cue complex. The task was to determine in which of the remaining two presentations the envelope of the cue band occurred. Though performance was impaired by the maskers, there was less interference than generally observed when modulated maskers are used in experiments evaluating AM detection and depth and rate discrimination (Yost, Sheft, and Opie 1989; Moore et al. 1991). The amount of interference did tend to increase with noise bandwidth. With the spectral location of the masking band having little effect, results indicated selective processing of the target band modulation, especially at the narrower noise bandwidths. When more than one masking band was present, there was little effect of coherence among the masking bands on the detection of target band synchrony. With masker coherence, the task then requires synchrony detection restricted to the spectral region of the target bands.

The second study by Sheft and Yost (1992a) evaluated the role of the carrier in modulation processing by measuring the ability to discriminate among patterns of modulation when the spectral location of the carrier was varied within a trial. Stimuli were again narrowband noises with the center frequency changed between the cue presentation and the two observation intervals of a single trial. To a varying extent, listeners were able to discriminate spectrally transposed modulation patterns, with this ability diminishing as the bandwidth of the noise bands, and consequently the average modulation rate, was increased. When multiple noise bands were used for either the cue or the observation intervals, listeners were unable to integrate modulation information across spectral regions. These results suggest that the modulation information was not processed independently of the spectral location of the carrier.

7.3 Cross-Spectral Processing

In the two studies of Sheft and Yost (1990, 1992a), the best performance was obtained at the narrower noise bandwidths where the average fluctuation rate of the noise bands is low. Several lines of evidence suggest slow AM is important in source determination. Coherent low-frequency modulation is a characteristic of many sound sources. Also, the temporal envelope of speech is dominated by periodicities of less than 20 Hz (Plomp 1983). Modeling of auditory temporal resolution as a low-pass process (e.g., Viemeister 1979) suggests greater fidelity in the coding of slower envelope modulation. Finally, most listening environments are reverberant. The effect of reverberation is to reduce the depth of high-frequency modulation and, in fact, enhance the transmission of low-frequency modulation information in the presence of background noise (Houtgast, Steeneken, and Plomp 1980).

Results from a variety of psychophysical paradigms demonstrate cross-spectral processing of a relatively slow envelope modulation. This work includes studies of comodulation masking release (Hall and Grose 1990; Moore, Glasberg, and Schooneveldt 1990), absolute detection of modulated waveforms (Cohen and Schubert 1987; McFadden 1987; Wright 1990), discrimination of envelope correlation (Richards 1987; Wakefield and Edwards 1987; Yost and Sheft 1989), modulation detection interference (Yost and Sheft 1989; Yost, Sheft, and Opie 1989; Hall and Grose 1991; Moore et al. 1991), and discrimination of stimuli based upon which components are modulated (Sheft and Yost 1989a,b, 1992b; Hall and Grose 1991). Investigators have interpreted some results in terms of spectral grouping based on common AM. The masked threshold for detecting a noise band (target) is higher if the target and masker bands share a common modulation. McFadden

(1987) has suggested that cross-spectral coherence of modulation groups the target and masker bands, making processing (in this case detection) of the individual bands more difficult. A similar argument has been applied by Yost and Sheft (1989) to modulation detection for a tonal carrier (probe) in the presence of a modulated masker. Modulation detection interference (MDI) is the increment in probe AM detection threshold due to masker modulation. However, results from MDI studies in which the probe and masker modulators vary in either rate or phase indicate greater interference than might be expected from grouping effects based on coherent modulation (Moore et al. 1991; Yost and Sheft 1989; Yost, Sheft, and Opie 1989). These results have been viewed by some as evidence against a grouping explanation for MDI (Moore et al. 1991; also see Viemeister and Plack, Chapter 4).

Hall and Grose (1990) extended the study of comodulation masking release (CMR) to conditions in which multiple modulation patterns allowed for auditory grouping. The basic CMR result (see Green, Chapter 2) is that the masked threshold of a tonal signal centered in a narrowband noise can be lowered by adding coherently modulated noise bands spectrally remote from the signal frequency. The addition of independently modulated (deviant) bands can then reduce the masking release. Hall and Grose (1990) found that the CMR could be restored by either increasing the number of deviant bands or introducing a gating asynchrony between the deviant bands and bands comodulated with the on-signal frequency masker. They interpreted the restored CMR to the perceptual segregation of the deviant bands from the on-signal band.

Though many of the results from studies involving cross-spectral modulation can be interpreted in terms of grouping effects based on modulation, these studies do not directly demonstrate this type of processing. Bregman and his coworkers (Bregman et al. 1985; Bregman, Levitan, and Liao 1990) studied the effect of AM on spectral grouping, evaluating relatively rapid envelope periodicities in the range of the glottal pulsation of vowel sounds. They found that the perceptual fusion of two spectrally distant carriers was affected by coherence of modulation. When the modulating waveforms of the two carriers differed in either frequency or phase, spectral fusion was diminished. Harmonicity among the carriers had no effect, indicating that segregation judgments were based upon cross-spectral processing of envelope periodicity.

Sheft and Yost (1989b, 1992b) evaluated the effect of AM on spectral grouping, but at low modulation rates. In these studies, the two stimuli presented on each trial of the 2IFC task were spectrally identical before modulation. The components of the target subset were then distinguished from the complex by modulation. Either cuing of the target subset or the residue pitch of the subset were used as an identifier of the target. Results indicated that coherent AM at rates from 5 to 25 Hz of harmonically related target components led to proper spectral grouping. With inharmonic components or cross-spectrally incoherent modulation, performance could be modeled by assuming the accumulation of independent sources of modulation information not requiring spectral grouping per se.

Speculation on spectral grouping based on envelope characteristics requires both carrier-specific coding of AM within the CANS and then cross-spectral processing sensitive to envelope relationships. Single units at each of the major levels of the mammalian central auditory nervous system encode AM (e.g., Møller 1976; Creutzfeldt, Hellweg, and Schreiner 1980; Schreiner and Urbas 1986; Rees and Palmer 1989; Kim, Sirianni, and Chang 1990). Similar to the coding of signal fine structure, the units exhibit a phase-locked response to the envelope of a modulated stimulus if the carrier is within the unit's excitatory region. Recent physiological studies indicate a topographical organization according to AM frequency within nuclei of the CANS (Schreiner and Langer 1988; Kim, Sirianni, and Chang 1990). Systematic variation across the nuclei in terms of selectivity to AM frequency represents a spatial mapping of temporal envelope information. Results suggest that the place mapping of envelope periodicity may lie orthogonally to the tonotopic organization of the nuclei. This two-dimensional mapping allows for the grouping of spectral components that share a common envelope periodicity.

Yost, Sheft, and Opie (1989) suggested two schemes for grouping spectral components that share a common form of modulation (see Fig. 6.18). Based on models presented by McFadden (1988) for residue pitch processing, both schemes are consistent with a two-dimensional neural organization that maps modulation frequency against carrier frequency. The first scheme relies on recognition of the response pattern within the two-dimensional array. The second scheme postulates AM channels whereby all units excited by a common modulation provide converging input to a higher neural center (see Kay 1982 for a discussion of modulation channels; also Viemeister and Plack, Chapter 4). This scheme thus requires neural processing above the mapping of the midbrain. Though temporal resolution (the encoding of high-frequency AM) diminishes at successively higher levels of the CANS, the gain in the response to low-frequency AM increases (Møller 1976; Creutzfeldt, Hellweg, and Schreiner 1980; Schreiner and Urbas 1986; Rees and Palmer 1989; Kim, Sirianni, and Chang 1990). For low modulation rates, the response of cortical units provides the best correlate to psychophysical results (Fastl et al. 1986). A third processing scheme for source segregation that does not rely on place mapping of envelope periodicity was proposed by Malsburg and Schneider (1986). In their model, spectral segregation is based on the recognition of the synchronized temporal response of units tuned to different carrier frequencies. Neural synchronization thus serves to represent coherent features of a cross-spectrally defined pattern.

All three models allow for spectral grouping based on common AM. The models, however, can be distinguished in terms of the cross-spectral correla-

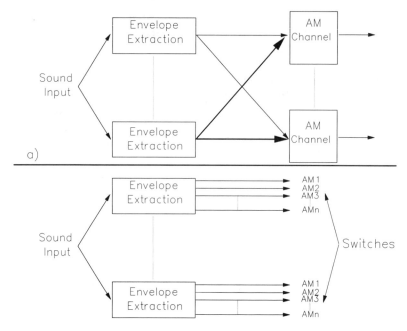

FIGURE 6.18. Two schematic diagrams indicating how modulation might be used to fuse and segregate spectral components for sound source determination. Envelope detection occurs for each peripheral auditory filter channel. In the top diagram, the outputs of each envelope extractor is fed to each AM channel. Only those spectral components that have envelope patterns to which the channel is tuned are passed and processed by that channel. In the bottom diagram, the output of each envelope extractor is switched to a path (AMi) based on its envelope pattern. Recognition of the pattern of excitation across the various paths allows for spectral grouping. (Based on similar diagrams by Yost, Sheft, and Opie 1989.)

tion of modulation information and the retention of the relationship between envelope and carrier characteristics. The pattern recognition schemes maintain a close association between envelope and carrier characteristics; spectral grouping based on AM channels does not. As described above, the measures of the ability to associate AM with its carrier depend on the procedure used. Results from the simplest paradigm in which the task is to discriminate complex stimuli based on which components are modulated show that a large AM depth is needed to achieve good discrimination. These results indicate poor associative ability. However, with more complex tasks of either synchrony detection or envelope pattern discrimination, modulation does not appear to be processed independently of the spectral location of the carrier. Thus, at present, experimental results do not rule out any of the three models for spectral grouping.

8. Methods and Models of Fusion and Segregation

Models of neural coding by the auditory periphery provide insights into how the auditory system might process the information in the neural code to allow for sound source determination. Models which provide a dynamic view of neural coding are particularly useful (Patterson 1987; Slaney and Lyons 1991; Patterson et al. 1992). Patterson suggested that there are three basic perceptions that are extracted from a complex sound generated by many sound sources: transients, periodic sounds, and noise. Patterson's model (Patterson et al. 1992) and that of Slaney and Lyons (1991) yielded similar "dynamic pictures" of auditory processing. These models have three crucial stages: 1) a filter bank which simulates the frequency selectivity performed by the biomechanics of the cochlea; 2) neural transduction which provides an adaptive, rectified, compressed, and sharpened (both temporally and spectrally) version of the filter outputs; and 3) a triggering mechanism which processes periodic waveforms. The first two stages capture the major dimensions of static sounds, while the third stage creates a dynamic pictorial representation of the perception of sound. Three types of sounds—transients, periodic sounds, and noise (long duration aperiodic sounds)—are clearly separable in these models, suggesting that these classes of sounds are important components for solving the task of sound source determination (see Patterson et al. 1992).

Sound source determination requires both analytic and synthetic processing. Analytically, a complex, time-varying sound field must be partitioned, with the concurrent spectral components and sequential acoustic events grouped according to their respective source. Synthetically, the components and events originating from a single source must be perceived as representing a single entity. Models for spectral fusion based on pitch and AM are described in Section 7.2. These models are neurally based, working on the peripheral output. Assmann and Summerfield (1990) and Meddis and Hewitt (1992) have utilized nonlinear neural processing preceding periodicity analysis to estimate the fundamentals of two concurrent vowels. Beauvois and Meddis (1991) have applied a peripherally based model to stream segregation. Thus, both fusion among concurrent spectral components and segregation of sequential auditory events can be modeled by a common processing framework.

These models demonstrate many of the key elements that must be preserved in the neural code to allow for sound source determination. However, most of the models have not yet provided an account for how this coded information is pooled or fused in order to form the image we hear. That is, the models do not suggest a mechanism for auditory fusion. By viewing the dynamic outputs of these models, one can "see" many of the images one hears, but this does not explain how the auditory system forms these images.

A common theme in many of the studies discussed in this chapter is the role of cross-spectral processing in sound source determination. That is, in a multisource environment, the spectral components of the various sounds are likely to be intermixed, and the auditory system must be able to process all of the components in order to fuse some components and segregate others. A number of models and schemes have been studied that suggest such cross-spectral processing. The various theories of complex pitch perception covered in this chapter and by Moore (Chapter 3) assume that the entire spectrum of the sound is processed before a pitch is determined. Likewise, most of the models of binaural processing mentioned in this chapter and by Wightman and Kistler (Chapter 5) analyze the entire spectrum of the sound before determining a spatial location for the sound. More recently, Berg's (1990) model of spectral weights (see Green, Chapter 2) provided a broadband spectral scheme to describe how different spectral components are used by a listener in processing a spectral profile.

Yost (Chapter 1, Table 1) outlined a series of four questions that are important for understanding auditory perception and sound source determination. A number of different lines of investigation have revealed some crucial aspects of the physical dimensions of a complex sound field that may aid the auditory system in determining the sources of sounds (question 1). A great deal is known about how many of these variables are coded within the auditory system (question 2). Some past and a number of more recent models and experimental designs have addressed the question of how the auditory system fuses and segregates the information in the complex sound field such that many sound sources can be determined (questions 3 and 4). However, the work in this area of auditory perception and psychoacoustics is in its infancy, and major new discoveries are surely just around the corner.

Acknowledgments. We would like to thank our colleagues at the Parmly Hearing Institute for discussing the topics of this chapter with us. We are also grateful for the support provided by the National Institute on Deafness and the Other Communication Disorders, the National Science Foundation, and the Air Force Office of Scientific Research.

References

Assmann PF, Summerfield Q (1990) Modeling the perception of concurrent vowels: Vowels with different fundamental frequencies. J Acoust Soc Am 88:680–697.

Bacon SP, Smith MA (1991) Spectral, intensive, and temporal factors influencing overshoot. Quart J Exp Psychol 43A:373–399.

Barsz K (1991) Auditory pattern perception: The effect of tone location on the discrimination of tonal sequences. Percept Psychophys 50:290–296.

Bashford JA, Warren RM (1987) Multiple phonemic restorations follow the rules for auditory induction. Percept Psychophys 42:114–121.

Beauvois MW, Meddis R (1991) A computer model of auditory stream segregation. Quart J Exp Psychol 43A:517–541.

Beerends JG, Houtsma AJM (1986) Pitch identification of simultaneous dichotic two-tone complexes. J Acoust Soc Am 80:1048–1057.

Beerends JG, Houtsma AJM (1989) Pitch identification of simultaneous diotic and dichotic two-tone complexes. J Acoust Soc Am 85:813–819.

Berg BG (1990) Observer efficiency and weights in a multiple observation task. J Acoust Soc Am 88:149–158.

Bernstein LR, Green DM (1988) Detection of changes in spectral shape: Uniform vs. non-uniform background spectra. Hear Res 34:157–166.

Boring EG (1942) Sensation and Perception in the History of Experimental Psychology. New York: Appleton-Century.

Bregman AS (1990) Auditory Scene Analysis. Cambridge, MA: MIT Press.

Bregman AS, Campbell J (1971) Primary auditory stream segregation and perception of order in rapid sequences of tones. J Exp Psychol 89:244–249.

Bregman AS, Doehring P (1984) Fusion of simultaneous tonal glides: The role of parallelness and simple frequency relations. Percept Psychophys 36:251–256.

Bregman AS, Pinker S (1978) Auditory streaming and the building of timbre. Canad J Psychol 32:19–31.

Bregman AS, Abramson J, Doehring P, Darwin CJ (1985) Spectral integration based upon common amplitude modulation. Percept Psychophys 37:483–493.

Bregman AS, Levitan R, Liao C (1990) Fusion of auditory components: Effects of the frequency of amplitude modulation. Percept Psychophys 47:68–73.

Bregman AS, Liao C, Levitan R (1990) Auditory grouping based on fundamental frequency and formant peak frequency. Canad J Psychol 44:400–413.

Broadbent DE, Ladefoged P (1957) On the fusion of sounds reaching different sense organs. J Acoust Soc Am 29:708–710.

Brokx JPL, Nooteboom SG (1982) Intonation and the perceptual separation of simultaneous voices. J Phon 10:23–36.

Buell TN, Hafter ER (1991) Combination of binaural information across frequency bands. J Acoust Soc Am 90:1894–1900.

Carlyon RP (1989) Changes in the masked thresholds of brief tones produced by prior bursts of noise. Hear Res 41:223–236.

Carlyon RP (1991) Discriminating between coherent and incoherent frequency modulation of complex tones. J Acoust Soc Am 89:329–340.

Carlyon RP, White LJ (1992) Effect of signal frequency and masker level on the frequency regions responsible for the overshoot effect. J Acoust Soc Am 91:1034–1041.

Chalikia MH, Bregman AS (1989) The perceptual segregation of simultaneous auditory signals: Pulse train segregation and vowel segregation. Percept Psychophys 46:487–496.

Champlin CA, McFadden D (1989) Reductions in overshoot following intense sound exposures. J Acoust Soc Am 85:2005–2011.

Cherry EC (1953) Some experiments on the recognition of speech, with one and with two ears. J Acoust Soc Am 25:975–979.

Ciocca V, Bregman AS (1987) Perceived continuity of gliding and steady-state tones through interrupting noise. Percept Psychophys 42:476–484.

Cohen MF, Schubert ED (1987) The effect of cross-spectrum correlation on the detectability of a noise band. J Acoust Soc Am 81:721–723.

Cramer EM, Huggins WH (1958) Creation of pitch through binaural interaction. J Acoust Soc Am 30:413–417.

Creutzfeldt 0, Hellweg FC, Schreiner C (1980) Thalamocortical transformation of responses to complex auditory stimuli. Exp Brain Res 39:87–104.

Dannenbring GL, Bregman AS (1976) Stream segregation and the illusion of overlap. J Exp Psychol: Human Percept Perf 2:544–555.

Dannenbring GL, Bregman AS (1978) Streaming vs. fusion of sinusoidal components of complex waves. Percept Psychophys 24:369–376.

Darwin CJ (1981) Perceptual grouping of speech components differing in fundamental frequency and onset time. Quart J Exp Psychol 33A:185–207.

Darwin CJ (1984) Perceiving vowels in the presence of another sound: Constraints on formant perception. J Acoust Soc Am 76:1636–1647.

Darwin CJ, Ciocca V (1992) Grouping in pitch perception: Effects of onset asynchrony and ear of presentation of a mistuned component. J Acoust Soc Am 91:3381–3391.

Darwin CJ, Gardner RB (1986) Mistuning a harmonic of a vowel: Grouping and phase effects on vowel quality. J Acoust Soc Am 79:838–845.

Darwin CJ, Sutherland NS (1984) Grouping frequency components of vowels: When is a harmonic not a harmonic? Quart J Exp Psych 36A:193–208.

de Boer E (1976) On the "residue" and auditory pitch perception. In: Keidel WD, Neff WD (eds) Handbook of Sensory Physiology, Volume V/1. New York: Springer-Verlag, pp. 479–583.

Demany L, Semal C (1990) The effect of vibrato on the recognition of masked vowels. Percept Psychophys 48:436–444.

Deutsch D (1980) Ear dominance and sequential interactions. J Acoust Soc Am 67:220–228.

Dye RH (1990) The combination of interaural information across frequencies: Lateralization on the basis of interaural delay. J Acoust Soc Am 88:2159–2170.

Fastl H, Hesse A, Schorer E, Urbas J, Müller-Preuss P (1986) Searching for neural correlates of the hearing sensation fluctuation strength in the auditory cortex of squirrel monkeys. Hear Res 23:199–203.

Gardner RB, Darwin CJ (1986) Grouping vowel harmonics by frequency modulation: Absence of effects on phonemic categorization. Percept Psychophys 40:183–187.

Gardner RB, Gaskill SA, Darwin CJ (1989) Perceptual grouping of formants with static and dynamic differences in fundamental frequency. J Acoust Soc Am 85:1329–1337.

Goldstein JL (1973) An optimum processor theory for the central formation of the pitch of complex tones. J Acoust Soc Am 54:1496–1516.

Green DM (1988) Profile Analysis. New York: Oxford Press.

Green DM, Yost WA (1975) Binaural analysis. In: Keidel WD, Neff WD (eds) Handbook of Sensory Physiology, Volume V/2. New York: Springer-Verlag, pp. 461–480.

Hall JW III, Grose JH (1990) Comodulation masking release and auditory grouping. J Acoust Soc Am 88:119–125.

Hall JW III, Grose JH (1991) Some effects of auditory grouping factors on modulation detection interference (MDI). J Acoust Soc Am 90:3028–3035.

Handel S (1989) Listening: An Introduction to the Perception of Auditory Events. Cambridge, MA: MIT Press.

Hartmann WM (1988) Pitch perception and the organization and integration of auditory entities. In: Edelman GW, Gall WE, Cowan WM (eds) Auditory Function: Neurobiological Bases of Hearing. New York: John Wiley and Sons, pp. 623–645.

Hartmann WM, McAdams S, Smith BK (1990) Hearing a mistuned harmonic in an otherwise periodic complex tone. J Acoust Soc Am 88:1712–1724.

Helmholtz H (1885) On the Sensations of Tone as a Physiological for the Theory of Music. New York: Dover Press (1954).

Hirsh IJ (1988) Auditory perception and speech. In: Atkinson RC, Herrnstein RJ, Lindzey G, Luce RD (eds) Steven's Handbook of Experimental Psychology, Volume 1. New York: John Wiley and Sons, pp. 377–408.

Houtgast T (1972) Psychophysical evidence for lateral inhibition in hearing. J Acoust Soc Am 51:1885.

Houtgast T, Steeneken HJM, Plomp R (1980) Predicting speech intelligibility in rooms from the modulation transfer function. I. General room acoustics. Acustica 46:60–72.

Houtsma AJM, Goldstein JL (1972) The central origin of the pitch of complex tones evidence from musical interval recognition. J Acoust Soc Am 51:520–529.

Judd T (1979) Comments on Deutsch's musical scale illusion. Percept Psychophys 26:85–92.

Kay RH (1982) Hearing of modulation in sounds. Physiologic Rev 62:894–975.

Kidd GR, Watson CS (1993) The proportion-of-the-total-duration rule for frequency resolution in ten-tone patterns. J Acoust Soc Am (in press).

Kidd GR, Watson CS (1992b) The proportion-of-the-total-duration (PTD) rule for holds for duration discrimination. J Acoust Soc Am 92:2318.

Kim DO, Sirianni JG, Chang, SO (1990) Responses of DCN-PVCN neurons and auditory nerve fibers in unanesthetized decerebrate cats to AM and pure tones: Analysis with autocorrelation/power spectrum. Hear Res 45:95–113.

Kubovy M (1987) Concurrent pitch segregation. In: Yost WA, Watson CS (eds) Auditory Processing of Complex Sounds. Hillsdale, NJ: Lawrence Erlbaum Associates, pp. 299–314.

Kubovy M, Daniel JE (1983) Pitch segregation by interaural phase momentary amplitude disparity and by monaural phase. J Audio Eng Soc 31:630–634.

Malsburg Ch von der, Schneider W (1986) A neural cocktail-party processor. Biol Cybernet 54:29–40.

Marin CMH, McAdams S (1991) Segregation of concurrent sounds. II: Effects of spectral envelope tracing, frequency modulation coherence, and frequency modulation width. J Acoust Soc Am 89:341–351.

McAdams S (1984) Spectral fusion, spectral parsing, and the formation of auditory images. Stanford, CA: Stanford University (unpublished dissertation).

McAdams S (1989) Segregation of concurrent sounds. I: Effects of frequency modulation coherence. J Acoust Soc Am 86:2148–2159.

McFadden D (1975) Masking and the binaural system. In: Tower DB (ed) The Nervous System, Volume 3. New York: Raven Press, pp. 137–146.

McFadden D (1987) Comodulation detection differences using noise-band signals. J Acoust Soc Am 81:1519–1527.

McFadden D (1988) Failure of a missing fundamental complex to interact with masked and unmasked pure tones at its fundamental frequency. Hear Res 32:23–39.

McFadden D (1989) Spectral differences in the ability of temporal gaps to reset the mechanisms underlying overshoot. J Acoust Soc Am 85:254–261.

McFadden D, Wright BA (1992) Temporal decline of masking and comodulation masking release. J Acoust Soc Am 92:144–156.

Meddis R, Hewitt MJ (1992) Modeling the identification of concurrent vowels with different fundamental frequencies. J Acoust Soc Am 91:233–245.

Miller GA, Heise GA (1950) The trill threshold. J Acoust Soc Am 22:637–638.

Miller JD (1984) Auditory processing of the acoustic patterns of speech. Arch Otolaryngol 110:154–159.

Møller AR (1976) Dynamic properties of primary auditory fibers compared with cells in the cochlear nucleus. Acta Physiol Scand 98:157–167.

Moore BCJ (1989) An Introduction to the Psychology of Hearing, Third Edition. New York: Academic Press.

Moore BCJ, Glasberg BR (1990) Frequency discrimination of complex tones with overlapping and non overlapping harmonics. J Acoust Soc Am 87:2163–2177.

Moore BCJ, Peters RW, Glasberg BR (1985) Thresholds for the detection of inharmonicity in complex tones. J Acoust Soc Am 77:1861–1867.

Moore BCJ, Glasberg BR, Schooneveldt GP (1990) Across-channel masking and comodulation masking release. J Acoust Soc Am 87:1683–1694.

Moore BCJ, Glasberg BR, Gaunt T, Child T (1991) Across-channel masking of changes in modulation depth for amplitude- and frequency-modulated signals. Quart J Exp Psych 43A:327–347.

Patterson RD (1987) A pulse ribbon model of monaural phase perception. J Acoust Soc Am 82:1560–1586.

Patterson RD, Robinson K, Holdsworth J, McKeown, Zhang C, Allerhand M (1992) Complex sounds and auditory images. In: Cazals Y, Demany L, Horner L (eds) Auditory Physiology and Perception. Oxford: Pergamon Press, pp. 429–443.

Plomp R (1976) Aspects of Tone Sensation. New York: Academic Press.

Plomp R (1983) The role of modulation in hearing. In: Klinke R, Hartmann R (eds) Hearing: Physiological Bases and Psychophysics. Berlin: Springer-Verlag, pp. 270–275.

Plomp R, Mimpen AM (1981) Effect of the orientation of the speaker's head and the azimuth of a noise source on the speech reception threshold for sentences. Acustica 48:325–328.

Pollack I (1975) Auditory informational masking. J Acoust Soc Am, Suppl 1 57:S5.

Price R (1955) A note on the envelope and phase-modulated components of narrow-band Gaussian noise. IRE Trans Inf Theory IT1:9–13.

Rasch RA (1978) The perception of simultaneous notes such as in polyphonic music. Acustica 40:21–33

Rayleigh L (1877) The Theory of Sound. New York: Dover Press, English Edition (1945)

Rees A, Palmer AR (1989) Neuronal responses to amplitude-modulated and pure-tone stimuli in the guinea pig inferior colliculus, and their modification by broadband noise. J Acoust Soc Am 85:1978–1994.

Rice SO (1954) Mathematical analysis of random noise. In Wax N (ed) Selected Papers on Noise and Stochastic Processes. New York: Dover Press, pp. 133–294.

Richards VM (1987) Monaural envelope correlation perception. J Acoust Soc Am 82:1621–1630.

Saberi K, Dostal L, Sadralodabai T, Bull V, Perrott DR (1991) Free-field release from masking. J Acoust Soc Am 90:1355–1370.

Scheffers MTM (1982) The role of pitch in the perceptual separation of simultaneous vowels II. IPO Annual Prog Rep 17:41–45.

Schreiner CE, Langer G (1988) Periodicity coding in the inferior colliculus of the cat. II. Topographical organization. J Neurophysiol 60:1823–1840.

Schreiner CE, Urbas JV (1986) Representation of amplitude modulation in the auditory cortex of the cat. I. The anterior auditory field (AAF). Hear Res 21:227–241.

Sheft S, Yost WA (1989a) Detection and recognition of amplitude modulation with tonal carriers. J Acoust Soc Am, Suppl 1 85:S121.

Sheft S, Yost WA (1989b) Spectral fusion based on coherence of amplitude n.odulation. J Acoust Soc Am, Suppl 1 86:S10–S11.

Sheft S, Yost WA (1990) Cued envelope-correlation detection. J Acoust Soc Am, Suppl 1 88:S145.

Sheft S, Yost WA (1992a) Spectral transposition of envelope modulation. J Acoust Soc Am 91:2333.

Sheft S, Yost WA (1992b) Concurrent pitch segregation based on AM. J Acoust Soc Am 92:2361.

Singh PG (1987) Perceptual organization of complex-tone sequences: A tradeoff between pitch and timbre? J Acoust Soc Am 82:886–899.

Slaney M, Lyon R (1991) Apple Hearing Demo Reel. Apple Technical Report #25. Cupertino, CA: Apple Computer, Inc.

Smith RL, Zwislocki JJ (1975) Short-term adaptation and incremental responses in single auditory-nerve fibers. Biol Cybernet 17:169–182.

Steiger H, Bregman AS (1982) Competition among auditory streaming, dichotic fusion, and diotic fusion. Percept Psychophys 32:153–162.

Stellmack MA, Dye RH (1993) The combination of interaural information across frequencies: The effects of number and spacing of components, onset asynchrony, and harmonicity. J Acoust Soc Am 93:2933–2947.

Summerfield Q, Assmann PF (1989) Auditory enhancement and the perception of concurrent vowels. Percept Psychophys 45:529–536.

Summerfield Q, Assmann PF (1991) Perception of concurrent vowel: Effects of harmonic misalignment and pitch-period asynchrony. J Acoust Soc Am 89:1364–1377.

Summerfield Q, Sidwell A, Nelson T (1987) Auditory enhancement of changes in spectral amplitude. J Acoust Soc Am 81:700–708.

Terhardt E, Stoll G, Seewann, M (1982) Pitch of complex signals according to virtual pitch theory: Test, examples, and predictions. J Acoust Soc Am 71:671–678.

Trahiotis C, Bernstein LR (1990) Detectability of interaural delays over select spectral regions: Effects of flanking noise. J Acoust Soc Am 87:810–813.

van Noorden LPAS (1975) Temporal coherence in the perception of tone sequences. Eindhoven: Eindhoven Universitat Technologie (unpublished dissertation).

van Noorden LPAS (1977) Minimum differences of level and frequency for perceptual fission of tone sequences ABAB. J Acoust Soc Am 61:1041–1045.

Verschuure J (1981) Pulsation patterns and nonlinearity of auditory tuning. I. Psychophysical results. Acustica 49:288–295.

Viemeister NF (1979) Temporal modulation transfer functions based upon modulation thresholds. J Acoust Soc Am 66:1364–1380.

Viemeister NF (1980) Adaptation of masking. In: van den Brink G, Bilsen FA (eds) Psychophysical, Physiological and Behavioral Studies in Hearing. Delft, the Netherlands: Delft University Press, pp. 190–198.

Viemeister NF, Bacon SP (1982) Forward masking by enhanced components in harmonic complexes. J Acoust Soc Am 71:1502–1507.

Wakefield GH, Edwards B (1987) Discrimination of envelope phase disparity. J Acoust Soc Am, Suppl 1 82:S41.

Wakefield GH, Edwards B (1989) Cross-spectral envelope phase discrimination for FM signals. J Acoust Soc Am, Suppl 1 85:S122.

Warren RM, Obusek CJ, Ackroff JM (1972) Auditory induction: Perceptual synthesis of absent sounds. Science 176:1149–1151.

Watson CS (1976) Auditory pattern discrimination. In: Hirsh SK, Eldredge DH, Hirsh IJ, Silverman SR (eds) Hearing and Davis. St. Louis, MO: Washington University Press, pp. 175–189.

Watson CS, Kelly WJ (1981) The role of stimulus uncertainty in the discrimination of auditory patterns. In: Getty DJ, Howard JH (eds) Auditory and Visual Pattern Recognition. Hillsdale NJ: Lawrence Erlbaum Associates, pp. 37–59.

Watson CS, Wroton HW, Kelly WJ, Benbassat CA (1975) Factors in the discrimination of tonal patterns. I. Component frequency, temporal position, and silent intervals. J Acoust Soc Am 57:1175–1185.

Wightman FL (1973a) Pitch and stimulus fine structure. J Acoust Soc Am 54:397–406.

Wightman FL (1973b) The pattern-transformation model of pitch. J Acoust Soc Am 54:407–417.

Wilson JP (1970) An auditory afterimage. In: Plomp R, Smoorenburg G (eds) Frequency Analysis and Periodicity Detection in Hearing. Leiden, The Netherlands: AW Sijthoff, pp. 303–312.

Woods WS, Colburn HS (1993) Test of a model of auditory object formation using intensity and ITD discrimination. J Acoust Soc Am (in press).

Wright BA (1990) Comodulation detection differences with multiple signal bands. J Acoust Soc Am 87:292–303.

Yost WA (1991) Auditory image perception and analysis: The basis for hearing. Hear Res 56:8–18.

Yost WA, Sheft S (1989) Across critical band processing of amplitude-modulated tones. J Acoust Soc Am 85:848–857.

Yost WA, Watson CS (1987) Auditory Processing of Complex Sounds. Hillsdale, NJ: Lawrence Erlbaum Associates.

Yost WA, Harder PJ, Dye RH (1987) Complex spectral patterns with interaural differences: Dichotic pitch and the "central spectrum." In: Yost WA, Watson CS (eds) Auditory Processing of Complex Sounds. Hillsdale, NJ: Lawrence Erlbaum Associates, pp. 190–201.

Yost WA, Sheft S, Opie, J (1989) Modulation interference in the detection and discrimination of amplitude modulation. J Acoust Soc Am 86:2138–2147.

Zwicker E (1964) Negative afterimage in hearing. J Acoust Soc Am 36:2413–2415.

Zwicker E (1965) Temporal effects in simultaneous masking an loudness. J Acoust Soc Am 38:132–141.

Zwicker UT (1984) Auditory recognition of diotic and dichotic vowel pairs. Speech Commun 3:265–277.

Index

Since the vast bulk of this volume deals with human data, humans are not indexed separately.

Absolute pitch, 94
Acoustics, in sound localization, 158ff
Adaptation
　neural, 212ff
　sound, 212ff
Amplitude modulation, 11, 78, 220ff, 224ff
　tuning, 129
Analytic listening, 200
Apparent sound position, *see* Sound source position
Asynchronous sources, 211ff
Audiogram, masked, 71, 73, 74
Auditory coding, 10–11
Auditory entity, 193ff
Auditory filter, 10–11, 57–58, 63, 64, 132ff
　and basilar membrane, 81
　and critical band, 81
　and level, 73
　and masking, 70ff
　and PTC, 61
　and time analysis, 132ff
　center frequency, 67–68
　characteristics, 67ff
　complex sounds, 75–77
　equivalent rectangular bandwidth, 68
　factors influencing, 75ff
　loudness, 75–77
　notched-noise filter, 83
　origin, 81–82

　outer and middle ear transfer function, 65–67
　pitch, 104
　relation to excitation pattern, 72–73
　sharpening, 81
　time course of development, 82–83
　variation, 67ff
Auditory filter shape, 58, 59ff
　and level, 69–70
　measurement, 67
　notched-noise method, 62–64
　PTC, 60–62
　rippled-noise method, 65
Auditory image, 193ff
Auditory object, 193ff
Auditory perception, 8–10, 193ff
　history, 195–196
Auditory periphery, and neural code, 3
Auditory processing, model, 229–230
Auditory scene, 193ff
Auditory sensitivity, 4
Autocorrelation, 204

Backward masking, 121
Basilar membrane, and auditory filter, 81
　and equivalent rectangular bandwidth, 67
　complex tones, 96–97
　sharpening, 81

237

Index

Binaural hearing, and head-related transfer function, 7
Binaural masking level difference, 207ff
Binaural pitch, 209

Central pitch, 201
CMR (Comodulation Masking Release), 33ff, 41ff, 38–139, 226–227
  amplitude modulation, 45–46
  psychometric functions, 42–43
  bandwidth, 45–46
  binaural, 43–44
  deferred, 41–42
  duration, 44
  frequency, 43
  frequency modulation, 46
  level, 44
  number of frequency bands, 44–45
  theory, 46ff
  uncertainty, 45
Cochlea
  and pitch, 95
  frequency selectivity, 81
  input spectrum, 65–67
Cochlear implant patients, time analysis, 34–135
Cocktail party effect, 205ff
Coherent modulation, 220ff, 225
Comodulation detection difference, 33–34, 49–50
  defined, 49
Comodulation masking release, see CMR
Complex pitch, 200, 203–204
Complex sounds, 2–4
  and auditory filter, 75–77
  and loudness, 75–77
  critical band, 77–78
  threshold, 77–78
Complex spectra, 197ff
Complex tones, 95–97
  and peripheral auditory system, 96–97
  audibility, 79–81
  basilar membrane, 96–97
  discrimination of pitch, 96
  inharmonic, 98–100
  pitch perception, 98

Correlation, envelope, 225
Critical band, 27, 56–59, 67, 68, 90–91
  and auditory filter, 81
  and equivalent rectangular bandwidth, 80–81
  auditory partials, 80–81
  complex sounds, 77–78
  loudness, 75–77
  pitch perception, 104
  time course, 82
Critical masking interval, 130ff, 144
Critical ratio, 58–59, 68
Cross correlation detector, 33
Cross-spectral correlation, 227ff
Cross-spectral processing, 225ff

Dichotic two-tone complexes, pitch, 101
Directional transfer function, 166, 167
Discrimination
  frequency, 90–93
  intensity, 4–5
DTF, see Directional transfer function
Duplex theory, 155, 158, 182–183

Encoding, see Neural encoding
Energy detector, 28, 32
Energy splatter, 29
Enhancement, auditory, 212ff
Envelope, defined, 117
Envelope detector, defined, 118
Envelope patterns, 116ff
Equal loudness contour, 67
Equalization and cancellation, 47
Equivalent rectangular bandwidth, 67, 70
  and basilar membrane, 67, 81
  and critical band, 80–81
  of auditory filter, 68
  variation with age, 68
ERB, see Equivalent rectangular bandwidth
Excitation pattern, 76–77
  and level, 73–76
  and masking, 70ff
  relation to auditory filter, 72–73

Fechner, 5
Fechner's Law, 26
Filter, auditory, 10–11, 57–58
Filter bank
   auditory, 229
   model, 196
Fine structure, time, 116ff
Fletcher, auditory filters, 56–57
Forward masking, 83–84, 85, 86, 121
Free field, 209
Frequency analysis, 56ff, 197
Frequency coding, 4
Frequency discrimination, 89ff, 90–93
   loudness cues, 91
   pure tones, 89ff
   temporal information, 93
   thresholds, 90, 92
Frequency modulation, 78, 220, 223ff
Frequency resolution, 203, *see also* Frequency analysis
Frequency selectivity, 86–88, *see also* Frequency analysis
Fusion, frequency components, 200–201

Gap detection, 120–121
Gaussian distribution function, 20
Gaussian noise, 21
Gestalt, 194–195

Harmonic, mistuned, 201–202
Harmonic series, best-fitting, 203–204
Harmonic sieve, 202–203
Harmonicity, 200ff
Head related transfer function, *see* HRTF
Hearing, 2–4
Hearing impairment
   gap detection, 134
   time analysis, 133ff
   TMTF, 134
Helmholtz, Herman, 4, 196, 224
Hilbert transform, 117
HRTF (Head Related Transfer Function), 7, 165, 167ff
   lateralization, 185–186
Huffman sequences, 138
Huggins pitch, 209

Informational masking, 217ff
Inharmonic complex tones, pitch, 98–100
Inhibition, neural, 83
Integration, temporal, 6
Integration time, intensity discrimination, 25
Intensity, 4–5
Intensity discrimination, 4–5, 13ff
   and loudness, 2, 26
   duration, 24
   frequency, 25, 26
   history, 13–14
   metric, 17ff
   noise, 20ff, 27ff
Intensity modulation, 131
Interaural level difference, 155ff, 209ff, 269, *see also* Sound localization
   acoustic determinants, 164
   bandwidth, 178, 180
   broadband sounds, 181–183
   effect of head shape, 159, 164
   frequency, 169–170
   head shadow, 177
   lateralization, 175
   measurements, 162–163
Interaural time difference, 155ff, 169, 209ff, *see also* Sound localization
   bandwidth, 178, 179
   broadband sounds, 181–183
   determined by HRTF, 168
   effect of head shape, 159, 164
   frequency, 169
   front–back confusion, 171
   lateralization, 175
   measurements, 159–161
Interval perception, 93–94

Lateral inhibition, 81
Lateral suppression, 81, 83ff
   and forward masking, 83–84
Lateralization, 7, 155, 156, 175, 209–210
Leaky integration, 118ff
Level, and excitation pattern, 73–76
   and pitch, 94–95
   auditory filter shape, 69–70

Index

Localization, 6–8, *see also* Sound localization
  cues, 6–7
  duplex theory, 7
  high frequencies, 7
  vertical, 7
Loudness, 59
  and complex sounds, 75–77
  and critical band, 75–77
  and intensity discrimination, 2, 26
  auditory filter, 75–77
  frequency discrimination, 91
  prediction, 76
  summation, 75

MAA, *see* Minimum audible angle
Mach, Ernst, 155
Masked audiogram, 71, 73, 74
Masker, noise, 59
Masker, notched-noise, 67
Masker, sinusoid, 59
Masking, 70ff, 209, 212, 224, 225
  across channel, 46
  adaptation, 139ff
  and auditory filter, 70ff
  and excitation pattern, 70ff
  assumptions, 57–58
  backward, 140ff
  binaural, 207ff
  forward, 139ff
  informational, 141, 207ff, 217ff
  nonsimultaneous, 139ff
  overshoot, 142–143, 212–213
  patterns, 70ff, 76
  period pattern, 125
  power spectrum model, 57–58, 59
  psychophysical tuning curve, 61
  repeated-gap technique, 84
  temporal, 139ff, 214–215
MDI (Modulation Detection Interference), 128ff, 139
Melody, 94
Middle ear, transfer function, 65–67, 69
Minimum audible angle, 173–174, 175
  paradigm, 179
Missing fundamental, 4, 95–96

Modulation, 78
  coherent, 220ff, 225
Modulation detection interference (MDI), 33ff, 50ff, 225ff
Modulation masking patterns, 129
Morse code, 4
Multiple look model, 146–147
Musical intervals, perception, 93–94

NBC chimes, 94
Negative afterimage, 213
Neural code, 2–4, 5, 119
  sound field, 8
Neural tuning curve, procedure, 60
Noise and signal detection, 29, 32
Noise
  intensity discrimination, 27ff
  masking, 57–58
Noise power density, defined, 27–28
Nonsimultaneous masking, 83, 84–85, 86–88
  nonadditivity, 141–142
Notched-noise masker, 61ff, 67, 83
Nowell identification, 203

Objective localization, 173–174
Off-time listening, 142
Offset asynchrony, 214–215
Ohm, acoustical law, 78
Onset asynchrony, 201, 214–215
Onset disparity threshold, 137–139
Outer ear, transfer function, 65–67
Overshoot, 142–143

Pattern perception, temporal, 218
PCA, *see* Principal Components Analysis
Perception
  auditory, 8–10
  musical intervals, 93–94
Periodicity and pitch, 203–204
Periodicity pitch, 96
Peripheral auditory system, analysis of complex tones, 96–97
Phase, effect on pitch, 103
Phase spectrum, 209

Phoneme identification, 223
Pinna, 156
  in HRTF, 168
  in sound localization, 165
  spectral cues, 185–186
  vertical localization, 180
Pitch, 4, 56ff, 88ff, 93–94, 200ff, 213
  absolute, 94
  auditory filter, 104
  central processing, 101
  critical band, 104
  dichotic two-tone complexes, 101
  effect of relative phase of components, 103
  matching, 202
  missing fundamental, 4
  of complex tones, 96
  of inharmonic complex tones, 98–100
  perception complex tones, 95–97
  perception models, 103ff, 203ff
  perception of pure tones, 88ff, 95
  spectral theory, 103–104
  temporal theory, 103–104
  theories, 88–89, 98
Place mechanism, and pitch, 94–95
Place theory, 4, 88–89, 90
Power spectrum model, and critical band, 56–59
Power-spectrum model, notched-noise method, 62–64
Principal components analysis, 168
Principle of dominance, 100–101
Profile analysis, 15, 16, 34ff, 197ff
  and frequency, 37ff
  and interstimulus interval, 36
  and phase, 37ff
  optimum decision rule, 40–41
  spectral weights, 39ff
  theory, 39ff
Psychoacoustics, definition, 1ff
Psychometric function, 20ff, 31ff
Psychophysical tuning curve, 60–62, 86–87
  auditory filter, 61
  masking, 61
  procedure, 60
Psychophysics, definition, 1
  development, 2

  in sound localization, 175
  procedures, 2
Pulsation threshold, 84–85
Pure tones, pitch perception, 88ff

Rayleigh, Lord, 14, 155, 195–196
Residue pitch, 96, 98ff, *see also* Pitch
Reverberation, 225
Riesz, Robert, 15, 16, 17, 136
Rippled noise, pitch, 101
Rippled spectrum, 198–199
Rippled-noise method, auditory filter shape, 65
ROEX filter, 63–64

Scaling, 2
Segregation, frequency components, 201
Segregation, perceptual, 226
Selective attention, 206–207
Sensation level and intensity discrimination, 25
Signal detection
  and duration, 29, 30
  and frequency, 30
  gated, 32
  in noise, 29, 32
  increment, 32
  optimum, 28–29, 33
  theory of, 2, 5
Signal to amplitude model, 87, 88
Simultaneous masking, 83, 84, 86–88
  lateral suppression, 83ff
Sound, description, 2
Sound localization, 2, 155ff, *see also* Interaural time difference, Interaural level difference, Localization
  acoustics, 158ff
  apparent sound position, 177ff
  bandwidth, 178
  broadband sounds, 181–183
  categorization, 176
  cues, 156
  distance, 155, 172
  duplex theory, 155, 158, 168
  dynamic cues, 171–172
  effects of head and torso, 165

Sound localization (cont.)
  effects of vision, 172
  elevation, 172–173
  frequency, 169–170
  front–back confusion, 171
  head movement, 171–172
  head shape, 158–159
  historical perspective, 155–158
  HRTF, 169
  individual differences, 186–187
  interaural differences, 168–170
  interaural level difference, 175
  interaural time difference, 175
  lateralization, 175, 184–185
  methods, 173ff
  minimum audible angle, 173–174, 179
  monaural, 170–171, 175, 176, 178, 182ff
  narrowband sounds, 178ff
  nonacoustical factors, 172–173
  objective vs subjective, 173–174
  outer ear filtering, 170
  pinna, 156, 165, 170, 180, 185–186
  principal components analysis, 168
  psychophysical procedures, 175
  pure tones, 179
  sound source position, 177ff
  spectral cues, 180
  stimulus delivery, 175
  vertical, 172–173, 179ff
  virtual sound sources, 184–186
Sound position
  determinants, 168ff
  interaural differences, 168–170
Sound source determination, 1ff, 193ff
Sound source segregation, 193ff, 223
Sound
  physics, 1–2
  temporal aspects, 5–6
Spectral fusion, 215, 229ff
Spectral grouping, 225ff
Spectral models for pitch, 203ff
Spectral notch, 213–214
Spectral pitch, 204
Spectral profile, 197–198
Spectral ripple, 213–214
Spectral segregation, 229ff
Spectral separation, 197

Spectral shape discrimination, *see* Profile analysis
Spectral theory, pitch perception, 103–104
Speech stimuli, 215
STAT model, *see* Signal to amplitude model
Stevens, S.S., scaling, 2
Stevens' Power Law, 26
Stream segregation, 212, 215ff, 223
Subjective localization, 173–174
Suppression, 85, 86, 214
Synchronous sources, 211ff
Synchrony
  onset, 214–215
  offset, 214–215
Synthetic listening, 200

Temporal acuity, defined, 118
Temporal analysis, brain, 227
Temporal aspects of sound, 5–6
Temporal integration, 6, 143ff
  detection and discrimination, 143–144
  energy detection, 145
  multiple looks, 146–147
  period, 141
  theory, 145ff
  time constant, 144
  vs resolution, 145–146
Temporal masking, 139ff
Temporal model for pitch, 203ff
Temporal modulation, 211ff
  coherent, 220ff
Temporal modulation transfer function, *see* TMTF
Temporal patterns, 116ff
Temporal resolution, 225, 226ff
  across channel, 137ff
  defined, 117ff
  level effects, 136–137
  limitations, 118ff
  vs integration, 145–146
Temporal separation, 211ff
Temporal theory, 89
  and pitch, 94
Temporal window, 130ff, 137
Theory of signal detectability, *see* Signal Detection

Threshold, complex sounds, 77–78
Threshold, frequency discrimination, 90, 92
Threshold, pulsation, 84–85
Timbre, 199–200, 213
Time analysis, 116ff
  across channel, 137ff
  and auditory filter, 132ff
  auditory filter bandwidth, 134ff
  filter center frequency, 132–133
  gap detection, 120–121
  hearing impairment, 133ff
  level effects, 136–137
  models, 121ff
  nonlinearities, 121
  systems analysis, 121ff
  time-reversed stimuli, 120
  within-channel, 120–121
TMTF, 123ff, 133
  and adaptation, 129
  and MDI, 128ff
  bandwidth, 128
  decision statistic, 127
  measurement, 124–125
  nonlinearities, 127–128
  theory, 125ff

Tonal residue, 100
Transfer function, outer and middle ear, 65–67, 69
Traveling wave, 4
TSD, 27, 28, *see also* Signal Detection
Two-tone suppression, 83, 84

Uncertainty, 29, 32, 218
Unmasking, 85–86

Ventriloquism, 172
Vibrato, 220
Virtual pitch, 96, 200–201, 202–203, 204
Virtual sound sources, in sound localization, 184–186
Vocal perception, 194–195, 199, 203
Vocal tract, 194–195, 203, 210
Vowel perception, 223

Weber, 5
Weber fraction, 18, 24, 27
Weber's Law, 5, 22ff, 136–137, 141
  near-miss, 22ff, 26, 136